Software Order Form

An Introduction to the Digital Analysis of Stationary Signals
Software forms an essential part of this *Computer Illustrated Text.*
It provides worked examples, calculating power where necessary and is a
great aid to understanding and learning.

Software is available for BBC Model B on 40/80 track $5\frac{1}{4}$ disc and for
IBM PC and compatibles on $5\frac{1}{4}$ disc at £6.00 per disc

To receive your software
   complete this form and return with your on or bef to:
   ESM, Duke Street, Wisbech, Cambs,PE13 2AE. Telephone 0945 83441
Telephone orders are accepted from customers **paying by credit card**

Please send me software for *An Introduction to the Digital Analysis of
Stationary Signals* by I P Castro on disc for

   ☐ BBC Model B 40/80 track   ☐ IBM PC

I enclose my cheque for £_____ payable to ESM

Please charge to my Access/Visa card number

☐☐☐☐☐☐☐☐☐☐☐☐☐☐☐☐   Expiry date _____

Name_____Signature _____

Address _____

_____

_____ **Date** _____

An Introduction to the Digital Analysis of Stationary Signals

Other titles in the series

An Introduction to Groups
D Asche

Introduction to Statistics
A W Bowman and D R Robinson

Regression and Analysis of Variance
A W Bowman and D R Robinson

Fourier Series and Transforms
R D Harding

A Simple Introduction to Numerical Analysis
R D Harding and D A Quinney

A Simple Introduction to Numerical Analysis Volume 2: Interpolation and Approximation
R D Harding and D A Quinney

From Number Theory to Secret Codes
T Jackson

Electric Circuit Theory
B E Riches

Introduction to Probability
D R Robinson and A W Bowman

A Computer Illustrated Text

An Introduction to the Digital Analysis of Stationary Signals

I P Castro

Department of Mechanical Engineering,
University of Surrey

Adam Hilger, Bristol and Philadelphia
ESM, Cambridge

British Library Cataloguing in Publication Data

Castro, I. P. (Ian P.)
  An introduction to the digital analysis
  of stationary signals.
  1. Telecommunication systems. Digital
  signals. Processing. Applications of
  microprocessor systems
  I. Title
  621.38'043

ISBN 0-85274-254-1 (Text)
ISBN 0-85274-255-X (Network pack)
ISBN 0-85274-256-8 (IBM disc)
ISBN 0-85274-257-6 (BBC 40/80 disc)

Library of Congress Cataloging-in-Publication Data are available

Series Editor: **R D Harding**, University of Cambridge

Published by: IOP Publishing Ltd
              Techno House, Redcliffe Way, Bristol BS1 6NX,
              England
              242 Cherry Street, Philadelphia, PA 19106, USA
              ESM, Duke Street, Wisbech, Cambs PE13 2AE,
              England
Published under the Adam Hilger/ESM imprint

Typeset by KEYTEC, Bridport, Dorset
Printed in Great Britain by WBC Print

D

621.38043

CAS

〉 Contents

▪ **Software Order Form**   **i**

Preface   **ix**

Acknowledgments   **xiii**

List of Software Packages   **xv**

1 **Background**   **1**
   1.1 Introduction   1
   1.2 The software package   2

2 **Examples of Stationary Signals**   **5**
   2.1 Introduction   5
   2.2 Simple periodic signals   5
   2.3 Almost periodic signals   6
   2.4 Random noise   8
   2.5 Gaussian random noise   9
   2.6 Noisy periodic signals   10
   2.7 Correlated noise   12
   2.8 Telegraph signal   14
   2.9 Software instructions   15

3 **Quantitative Description of Signal Content**   **17**
   3.1 Amplitude-domain statistics   18
   3.2 Time-domain statistics   33
   3.3 Further statistics   45
   3.4 Software instructions   46

4 Digital Sampling Criteria: Amplitude-domain Statistics **48**
  4.1 Introduction 48
  4.2 Quantisation and ranging errors 49
  4.3 Finite sample size errors 58
  4.4 The effects of using correlated samples 71
  4.5 Summary and examples 74
  4.6 Computer exercises 76

5 Digital Sampling Criteria: Time-domain Statistics **84**
  5.1 Introduction 84
  5.2 Aliasing 85
  5.3 Autocorrelation estimation 88
  5.4 Estimation of the spectral density function 101
  5.5 Windowing 106
  5.6 Summary and examples 110
  5.7 Computer exercises 115

6 Sample Laboratory Experiments **124**
  6.1 Introduction 124
  6.2 Typical laboratory sheets 124

Appendix A The Simulation of Random Signals **133**
  A.1 Introduction 133
  A.2 Generation of random noise (DAT0N8) 135
  A.3 Generation of Gaussian (white) noise (DAT1N8) 136
  A.4 Generation of correlated (pink) noise (DAT2N8) 137
  A.5 Generation of a second-order process (DAT3N8) 139

Appendix B Running the Software **141**
  B.1 Standard features 141
  B.2 Screen dumps 143

References **145**

Bibliography **146**

Index **147**

⟩ Preface

There are numerous books, including many now classical texts, on digital signal processing. In teaching final-year mechanical engineering students and postgraduate students with various backgrounds, however, it has in my experience been very difficult to find the kind of book that such students need as an introductory text. I believe that the teaching of signal processing should entail discussion of both the amplitude-domain statistics (such as signal means and variances) and the time-domain statistics (such as autocorrelation and spectra). Very few texts cover these two aspects of the subject together.

On the one hand, there are books on statistical probability theory, usually written by statisticians and often not with the needs of the engineering community in mind. On the other hand, there are many books on correlation and spectral analysis. In both cases, such books tend to be mathematically quite daunting and, even if written in the context of digital signal processing, the underlying physical principles tend to become buried in the analysis. For the mathematician, this may not matter very much but for the engineer, who is increasingly prone to attempt measurements of signal characteristics using a computer rather than buying a specific bench-top instrument, a proper understanding of the basic concepts can be vital.

Furthermore, I have found that concepts such as those embodied in the definitions of, for example, the probability density or autocorrelation functions are more easily grasped by students if they are given some opportunity to measure such quantities in the laboratory. There are no texts which attempt to impart the basic concepts in the context of direct experimentation.

This book, together with its associated software, has grown out of signal-processing courses developed over a number of years, in which the author has deliberately set out to back up the lecture content with practical experiments. Although the experiments are conducted on simulated signals originally generated entirely within the computer (with one or two exceptions), students usually have little difficulty in relating the different data sets to those which they would obtain by digitising real signals. This is almost certainly because the signal-processing classes have normally been part of much broader instru-mentation and microcomputing courses, within which the students *do* obtain practical experience in the use of analogue-to-digital conver-ters and computer interfaces in 'capturing' real signals.

Although in the author's courses at the University of Surrey the emphasis has generally been on the overall qualitative concepts rather than on the intricate mathematical detail, it is important that the student develops an ability to express the concepts symbolically and to undertake simple pieces of theoretical analysis. The level of mathematics used in this book is neither broader nor deeper than what is usually covered in the first two years of undergraduate mechanical engineering degree courses. It could not be less than that for a proper introduction to the subject and so, to that extent, the text is likely to be most appropriate for final-year undergraduate or postgraduate students. Furthermore, there is rather more material here than could normally be covered in, say, a 15–20 h 'signal-processing module' of a broader instrumentation and microcomputing course. The intention, however, has been to give sufficient develop-ment of the material within each subject area to provide both useful further reading for the student and some element of choice for the teacher in expanding his basic course.

After some introductory remarks, a brief description of the philoso-phy behind the software design is given in Chapter 1. Chapter 2 begins with a qualitative description of typical signals. Attention is concentrated, as the book's title implies, on stationary signals; stationarity is therefore defined in Chapter 2. This limitation might in some contexts be viewed as serious, but we believe that, until the student appreciates the concepts and problems which arise in analysis of stationary signals, he is unlikely to be able to grapple with the much more difficult problems arising from non-stationarity.

Chapter 3 discusses the various ways of characterising stationary signals, with the content organised largely in terms of amplitude-domain statistics (§3.1) and time-domain statistics (§3.2). The soft-ware accompanying Chapters 2 and 3 should be viewed as a useful visual aid to the text. Consideration of the sampling criteria necessary for adequate measurement of the signal characteristics forms the content of Chapters 4 and 5. Here, the software not only provides immediate illustrations of the text but also can be used to design a wide range of experiments on the simulated signals. An example of the latter is provided in Chapter 6.

The enterprising reader could go even further and generate quite different kinds of signal (either by digitising real signals or by concocting further simulated signals) and then use the software directly to measure the characteristics of the signals or as a basis for further laboratory classes.

If even a few readers are able to use both the text and the software as a basis for their own courses in digital signal processing, or even simply as an aid to their own understanding or that of their students, the effort involved in producing this further volume in the Computer Illustrated Text series will have been amply justified.

I P Castro
Spring 1988

⟩ Acknowledgments

My interest in signal processing has been stimulated and my understanding considerably enhanced by the seminal influence of Professor L J S Bradbury over many years. This book would never have appeared were it not for his continual encouragement and I am deeply grateful for the inspiration which he has provided.

The patient help of the secretarial staff of the Mechanical Engineering Department, University of Surrey, is also gratefully acknowledged. They are probably relieved that it is all over.

We are also grateful to Structured Software (15 Athelston Close, Bromborough, Wirral, Merseyside L62 2EX) for the use of a modified form of their Fast Fourier Transform routine in the BBC version of the software package which accompanies this book.

⟩ List of Software Packages

Display of the index   Chapter 1

Simple demonstration of each of these signals with   Chapter 2
options to display some user-generated variations
of them

Display of the ideal probability, autocorrelation   Chapter 3
and spectral functions for each of the signals in
Chapter 2

Demonstration of confidence limits for mean   §§4.1–4.5
values. User experiments on the measurement of
mean and mean-square values, higher-order
moments and probability densities. Demonstration
of the influence of time structure on the results
and user experiments on correlated data

Graphical display of any of the simulated signals   §4.6

Demonstration of aliasing effects   §5.2

User experiments on the measurement of spectral   §§5.3, 5.4
density and autocorrelation functions

〉 Chapter 1

〉 Background

〉1.1 Introduction

The measurement of physical variables such as temperature, fluid velocity, strain or vibration amplitudes and accelerations is a daily requirement in most fields of technology. Digitising the analogue output of a suitable transducer and performing calculations on the resulting sample values is commonplace and numerous essentially digital bench-top instruments (such as voltmeters and spectrum analysers) have been available for many years. The advantages of minimising the analogue content of a measurement system are many and obvious but one of the major disadvantages of bench-top instruments is their general lack of flexibility—one cannot normally measure a mean voltage or a probability distribution with a spectrum analyser, for example. They are also usually rather expensive when compared with the cost of an equivalent analogue-to-digital converter (ADC) interfaced to a small microcomputer.

With the rapid increase in the 'power-to-cost' ratio of microcomputer systems over recent years, it is therefore also becoming commonplace to transfer the digitised sample values directly to a computer, which is often used to control the digitising process in addition to performing the calculations necessary for particular measurements. Sometimes these calculations are done in 'real time', with updates of signal characteristics being obtained immediately after *each* digitised value is read. In other cases, particularly if rapid sampling rates are required, the calculations are done after a suitable block of samples has been obtained. In either case the digitised samples can be stored for later manipulation but, even if this is not done, it is clear that such a system is extremely flexible. Given a suitably rapid ADC and

1

computer interface, the kind of measurement that can be made is limited only by the 'ingenuity' of the software and the ability of the computer to perform repetitive arithmetic operations very quickly—computers are particularly good at the latter! In many ways, the computer is therefore the ideal 'instrument' to use for signal processing and, as in many quite different areas of technological endeavour, a major part of the effort lies in the design of suitable software.

This book is different from most others on digital signal processing in that it attempts not only to outline the fundamental ideas concerning the different ways in which a random signal can be usefully described but also to demonstrate how these ideas can be powerfully illustrated on even quite small computers. Although the text can be used as a reference and studied quite independently of the accompanying software package, it is likely to be even more useful when used in conjunction with this software. Some of the latter was originally developed as part of other packages used in a laboratory for the measurement of real signals, but the package accompanying this text has been designed essentially with the teaching function in mind and should be judged in that light. In §1.2 the major features of the software are outlined. Specific instructions for loading it and getting started (in the BBC context) are given in Appendix B. The reader using this book simply as a 'stand-alone' reference text could omit §1.2 and move straight to Chapter 2.

〉1.2   The software package

All the software was originally written for the standard BBC B microcomputer; the comments in the following paragraphs are therefore specific to the BBC context. For the PC version of the software, there were obviously far fewer memory constraints, which led to rather greater flexibility in the possible size of the data sets and the nature of the graphical displays.

Many of the calculations necessary for the measurement of signal characteristics can be time consuming, particularly when programmed in an interpretive language such as BASIC. Consequently the decision was taken prior to any of the software development that all the 'number-crunching' parts of the package should be written at assembler level, in the case of BBC machines, or a high-level compiled language in the case of PCs and compatibles. In the former case, to calculate signal variances, higher-order moments or autocorrelations, for example, this involved writing various multi-byte multiplication routines and embedding them in larger segments of assembler code so that all the arithmetic-intensive elements of the calculations are performed at machine code level. The outer BASIC routines simply supply the required number of data samples, the sampling rate, etc, to these machine code routines and then process the results as required.

It was felt that about 20 000 data values would provide a very adequate representation of most types of signal; use of memory space from &3000 to &7BFF (i.e. up to the screen mode 7 memory area in a standard BBC without shadow random-access memory) would allow 19 456 consecutive integer values of 1 byte and enough memory below &3000 for the necessary BASIC and machine code routines. This arrangement meant that all the data analysis has to be done *prior* to any use of modes 1 or 4 for graphical presentation, since these normally require considerable screen memory below &7BFF. The minimal disc access time required to load in a complete data set is an insignificant price to pay for the availability of such large data sets, although this does prevent use of cassette or tape filing systems, since these are unacceptably slow. Mode 1 allows colour graphics to be used, which considerably enhances the visual impact of the various displays. More recent machines of course have significantly greater memory capacities and it would be straightforward to modify the software to allow higher-quality graphics or larger data sets, or both. Note, however, that the software can be run on the MASTER series of BBC machines without any modification.

The data sets themselves were generated quite separately (see §4.6.1 and Appendix A for complete details). They are called into memory from the system disc by the software package as required. The latter is entirely menu driven and is quite large. In total the program code is over 200 kbytes long and is therefore divided into appropriate segments, largely coinciding with the particular chapters or sections of the text of the book, and loaded into core as required. Appendix B gives details of the contents of the disc and complete running instructions for the BBC version of the software. PC-based versions of the software have corresponding sets of instructions on files saved on the disc itself.

It should be emphasised again that integration between text and software is a little different in this volume than in some of the other

Computer Illustrated Texts. Chapters 2 and 3 contain essentially a qualitative discussion, definitions of particular characteristic functions and examples of the latter for a range of signals. The software associated with these chapters therefore embodies mainly a series of set illustrations—some of which are used as figures in the text. In a lecture course these displays could be used 'off line' (without the computer), but other displays are 'dynamic', in the sense that the computer is used essentially as an oscilloscope to show a continuous time display of particular kinds of signals. For the greatest impact, therefore, the computer should be used in conjunction with the material in the text. Still other options in the software associated with these chapters allow the user to generate his or her own signal by specifying appropriate parameters, and these also are clearly best demonstrated using the computer in the class (or in conjunction with reading the text). The final sections in Chapters 2 and 3 describe the various software options in detail.

Chapter 4 and 5 discuss the methods of actually measuring the signal characteristics and, again, it has been felt best to relegate the detailed description of the software options to final sections of their own, within each chapter. This minimises possible fragmentation of the earlier sections and allows the book to stand alone as an introduction to signal processing without the necessity for using the software. However, the routines available to support the text in these chapters give the freedom to measure any of the fundamental signal statistics using wide ranges of the appropriate parameters (such as the sampling rate and the number of samples for a mean value calculation). Consequently, these parts of the software are likely to be the most useful. With simple changes to the BASIC programs, they could even be used (and have been by the author and some of his postgraduate students) in a much more general context for the analysis of real data from the laboratory.

〉 Chapter 2

〉 Examples of Stationary Signals

〉 2.1 Introduction

In Chapter 3 we shall discuss quantitatively the basic descriptive properties of a stationary signal. Here our purpose is simply to provide illustrations of typical signals that occur in a wide range of physical situations and to discuss briefly the qualitative differences between them. The reader is encouraged to glance rapidly through the following sections, to read the software instructions in §2.9 and then to reread the text more thoroughly in conjunction with running the Chapter 2 package on the computer. All the illustrations in this chapter have been produced as screen dumps of the various displays produced by the software (but note that in Chapters 2 and 3 this is *not* possible using the screen dump option described in Appendix B). The appropriate figure number is included in the heading of each section.

〉 2.2 Simple periodic signals (figure 2.1)

This is one of the simplest kinds of completely deterministic signal. If the frequency and peak-to-peak amplitude are known, then the signal amplitude at any time $t > t_0$ in the future can be predicted with certainty from a known amplitude at $t = t_0$. Very few signals obtained from transducers would have this ideal form; there will almost always be some superimposed random noise and/or some frequency modulation arising either from the nature of the physical process being monitored or from the imperfect response of the transducer, or from both.

5

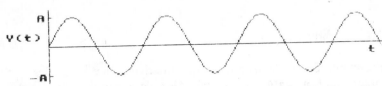

CHAPTER 2. STATIONARY SIGNALS - EXAMPLES
****

Cursor keys to select, RETURN for
demonstrations or ESCAPE for menu

Periodic signal,   'Almost' periodic
'Noisy-periodic'   Random noise
Gaussian noise   Correlated noise
Telegraph signal

This is a simple periodic signal,
defined by:

$$Y(t) = A*Sin(2\pi ft)$$

Figure 2.1. This and subsequent figures in Chapter 2 are all reproductions of screen displays corresponding to the highlighted signal. Thus, here, a simple periodic signal is shown.

⟩2.3 Almost periodic signals (figure 2.2)

We have shown here the ideal form of a signal which arises from a process containing two separate periodic components for which the ratio of the two frequencies is not a rational number. It can be represented by the simple linear sum of its two components. The case shown is

$$y(t) = A \sin(2\pi ft) + A \sin(2\sqrt{3}\pi ft).$$

Note that, since the two frequencies (f and $\sqrt{3}f$) are not rational multiples of one another, the fundamental period of the signal is infinite. It therefore never repeats itself but is still completely deterministic in the sense that the amplitude at any future time can be predicted if A and f are known.

Many physical cases of course are more complex than that illustrated here in that they contain many more frequency components. The usual way of expressing the $y(t)$ relationship is by using the Fourier series concept:

$$y(t) = A_0 + \sum_{n=1}^{\infty} A_n \cos(2\pi f_n t - \theta_n).$$

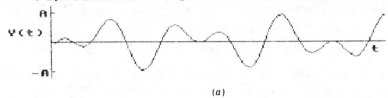

```
CHAPTER 2. STATIONARY SIGNALS - EXAMPLES
****************************************
```
```
Cursor keys to select, RETURN for
demonstrations or ESCAPE for menu
```
```
Periodic signal      'Almost-periodic'
'Noisy-periodic'     Random noise
Gaussian noise       Correlated noise
Telegraph signal
```
```
This is a signal defined by:

   Y(t) = A*Sin(2πft) + A*Sin(2√(3)πft)

The signal never repeats itself.
```
```
If, in contrast, the two contributing
frequencies were rationally related, the
signal would then have a finite period
- representable by a Fourier series.
```

(a)

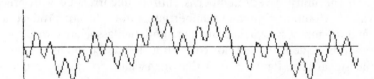

```
CHAPTER 2. STATIONARY SIGNALS - EXAMPLES
****************************************
```
```
Cursor keys to select, RETURN for
demonstrations or ESCAPE for menu
```
```
Periodic signal      'Almost-periodic'
'Noisy-periodic'     Random noise
Gaussian noise       Correlated noise
Telegraph signal
```
```
Generate your own complex wave
defined by:-

   y(t) = Sin(2πft) +

1.0Sin(2π4.0ft+180)+1.0Sin(2π.25ft+30_)
```
```
Type RETURN or SHIFT for another
```

(b)

Figure 2.2. (a) An almost periodic signal having an infinite period. (b) An almost periodic signal having three sinusoidal components with frequencies in the ratio 1:4:16. The signal repeats with a period of $4/f$.

In the example above, f_n/f_{n+1} is not rational, but very often there is a direct relationship so that

$$y(t) = A_0 + \sum_{n=1}^{\infty} A_n \cos(2\pi n f_1 t - \theta_n)$$

where f_1 is the 'fundamental' frequency, denoting the number of times that the signal repeats itself per unit time. Often only a few components are present. Almost never do physical phenomena produce simple sinusoidal data with only one component; even the output from a sine-wave generator, for example, will contain small contributions from higher-harmonic components. The reader can obtain some 'feel' for the typical waveform that results from the addition of a number of sinusoidal components by using the further option in the software package (by repeating the RETURN key press). This allows the display of signals described by

$$y(t) = \sin(2\pi f t) + A \sin(2\pi f_1 t + \theta_1) + B \sin(2\pi f_2 t + \theta_2)$$

in which the amplitudes A and B, frequencies f_1 and f_2 and phase angles θ_1 and θ_2 are arbitrary and must be defined by the user. To ensure that the total amplitude does not become too large for plotting, A and B should not exceed unity. In the example shown in figure 2.2(b), $f_1 = 4f$, $f_2 = f/4$, $\theta_1 = 180°$ and $\theta_2 = 30°$.

$\rangle$2.4   Random noise (figure 2.3)

This is a signal defined simply by

$$y(t) = 2A[0.5 - \text{RND}(1)_t]$$

where, at each t, $\text{RND}(1)$ is a nine-digit decimal number between zero and unity. Every number is entirely uncorrelated with preceding values. Computer random-number generators do not produce a truly random-number sequence but the subtleties associated with 'pseudo'-random sequences need not concern us here. The point to note about the signal (see figure 2.3) is that high values occur as frequently as low values, because the random-number generator produces every possible value with equal probability.

Assume that an infinite set of such sample sequences were available from the original signal. If the values obtained from each sequence at time t yield amplitude-domain statistics (mean, mean square, etc) independent of t and if in addition the 'sequence-averaged' correlation between samples at t and $t + \Delta t$ were also independent of t, the signal is said to be 'stationary'. Further, if the amplitude and correlation statistics obtained using the samples of one particular sequence did not depend on the sequence chosen, then the signal is also said to be 'ergodic'. Only stationary signals can be ergodic and the importance of an ergodic signal lies in the fact that all its properties can be determined from a single (suitably long) sample sequence. It is fortunate that, in practice, random data representing stationary physical phenomena are usually ergodic.

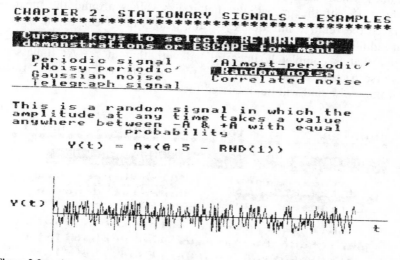

Figure 2.3. A random noise signal having a uniform amplitude probability density distribution (see §3.1.1) and the spectral characteristics of white noise (see §3.2.2).

Whilst these definitions are usually, as here, given in the context of random data, they also apply to periodic data. Provided that each sample sequence starts at random times (so that the signal phases will be random) a periodic signal is necessarily ergodic.

⟩2.5 Gaussian random noise (figure 2.4)

Unlike the previous random-noise case, the amplitudes here do not occur with equal probabilities. It is obvious simply by inspecting the signal that the likelihood of peak values is relatively lower than the likelihood of values near zero (compare figures 2.3 and 2.4). In fact, the amplitude probability function which is the basic measure of the amplitude-domain statistics (see Chapter 3) has in this case the well known Gaussian (or normal) form, defined by

$$p(y) = (1/\sigma\sqrt{2\pi})\exp[-(y - \bar{y})^2/\sigma^2]$$

where $\bar{y}$ is the mean value (zero in the case shown) and σ is the standard deviation—a measure of the variability of $y(t)$ about its mean value (see Chapter 3). Note that the probability that $y(t)$ has values which differ from the mean by more than about $\pm 3\sigma$ is only about 1% of the probability that it has values around the mean—hence the relative infrequency of the peak values seen in the signal. In addition, the signal shown here has, for all times t, an amplitude at t that is completely uncorrelated with that at $t + \Delta t$ (unless of course Δt is identically zero). This type of signal is commonly termed 'white' noise and will be discussed further in due course.

Figure 2.4. A random white noise signal having a Gaussian amplitude probability distribution (see §3.1.1) and a uniform spectral density distribution (see §3.2.2). Note the relatively rare occurrence of extreme values, compared with the signal in figure 2.3.

〉**2.6 Noisy periodic signals (figure 2.5)**

Here (figure 2.5(a)) we show a signal defined by

$$y(t) = A\sin(2\pi ft) + B[0.5 - \mathrm{RND}(1)_t].$$

CHAPTER 2. STATIONARY SIGNALS – EXAMPLES
****

Cursor keys to select, RETURN for
demonstrations or ESCAPE for menu

Periodic signal   'Almost-periodic'
Noisy-periodic   Random noise
Gaussian noise   Correlated noise
Telegraph signal

This is a signal characteristic of a
dominant periodic process affected by
uncorrelated random noise.

$$Y(t) = A*Sin(2\pi ft) + B*(0.5-RND(1))$$

$$B/A = 5/4 \text{ in this case}$$

Y(t)

(a)

CHAPTER 2. STATIONARY SIGNALS – EXAMPLES
****

Cursor keys to select, RETURN for
demonstrations or ESCAPE for menu

Periodic signal   'Almost-periodic'
Noisy-periodic   Random noise
Gaussian noise   Correlated noise
Telegraph signal

Generate your own noisy sin wave
defined by:-

$$y(t) = Sin(2\pi ft) + 3.0.(0.5-RND(1))$$

Type RETURN or SPACE for another

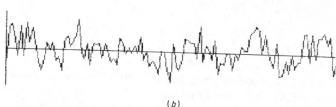

(b)

Figure 2.5. (a) A periodic signal contaminated by random noise whose amplitude is $\frac{5}{4}$ times the sine-wave amplitude. (b) As in (a), but with the noise amplitude three times the sine-wave amplitude. Note that the underlying periodicity is now very difficult to discern visually.

There is an underlying periodic component given by the first term, but it is 'hidden' by the addition of random noise of the sort described in §2.4. The degree to which the basic periodicity is submerged in the noise is clearly a function solely of the amplitude ratio B/A. In the case shown in figure 2.5(a), $B/A = \frac{5}{4}$ and the periodicity is still quite evident. Note, however, that whatever the value of B/A the signal is not deterministic. (Recall that we use this term simply to denote whether or not signal *amplitudes* at future times can be predicted from those in the past.)

The effect of varying the signal-to-noise ratio can be demonstrated by using the additional option in the software package (press the RETURN key again). This allows the user to display signals with any value for B/A. The example shown in figure 2.5(b) has $B/A = 3$; at a first glance, the underlying periodicity is not very noticeable. In later chapters, it will be demonstrated that even very noisy periodic signals can be recovered by special signal analysis techniques.

⟩2.7 Correlated noise (figure 2.6)

Visual comparison of this signal with that discussed in §2.5 shows immediately that, although peak values are relatively infrequent as before, there is a definite 'time structure' to the signal. Any instantaneous amplitude is *most* likely to be close to the immediately preceding values. This is a result of the fact that for this signal the correlation between two values $y(t)$ and $y(t + \tau)$, averaged over all values of t, is non-zero. Mathematically the autocorrelation $R(\tau)$ is defined by (for a signal $y(t)$ having zero mean)

$$R(\tau) = \lim_{T \to \infty} \left(\frac{1}{T} \int_0^T y(t)y(t + \tau)\,dt \right) \qquad \tau < \infty.$$

The signal shown here has $R(\tau) = \exp(-\tau)$.

For the random signals described in §§2.4 and 2.5, $R(\tau) = 0$ except for $\tau = 0$ (when $R(\tau)$ becomes simply the mean square). Obviously, any signal with one or more periodic components (e.g. those in §§2.2, 2.3 and 2.6) also has structure in time. Very many signals arising from physical phenomena, even if they have no *dominant* periodicity, will have a specific $R(\tau)$ function which only tends to zero as $\tau \to \infty$. This implies, in fact, the presence of a complete spectrum of Fourier (sinusoidal) components. The link between $R(\tau)$ and the signal spectrum—expressing the way in which the energy in the signal is spread throughout the frequency range—will be developed later. One should note, however, that even the autocorrelation function for random or Gaussian noise $(R(\tau \neq 0) = 0)$ implies a certain spectrum; in fact in both these cases the energy is spread equally over all frequencies, even though the amplitude statistics are quite different. (This *can* make true comparisons on a visual display unit screen rather misleading, since only discrete data points are plotted, with lines drawn between them. So, if the random noise signal is 'stretched' in the t direction too much when plotted, it will appear to have structure in time.)

Figure 2.6. Gaussian noise having an exponentially decaying autocorrelation function (see §3.2.1). This is known as pink noise. Note that these data and those shown in figure 2.4 are segments of signals which have *identical* probability density functions but, obviously, quite different *time*-domain statistics.

It cannot be overemphasised that two signals with identical amplitude-domain statistics do not necessarily have the same time structure (spectra). This is perhaps most clearly evident by considering the sine wave. If parts of it were chopped out and placed somewhere else on the time axis, the amplitude-domain statistics (mean, mean square, etc) would not change although the time structure clearly would. Equally important is the fact that two signals having identical structure in time, denoted by the same autocorrelation (or spectrum) function, may have quite different amplitude-domain statistics. This is demonstrated rather starkly by the next sample signal.

〉2.8   Telegraph signal (figure 2.7)

In this signal, the time intervals between each change in amplitude
from high to low (or vice versa) are distributed according to a
Poisson function. It can be shown that in those circumstances the
autocorrelation is defined by $R(\tau) = \exp(-\tau)$ (Snyder 1975); this is
an identical time structure to that of the correlated noise described in
§2.7. Energy is therefore distributed across the frequency range in
just the same way, although one might not deduce that by visual
inspection of the signal!

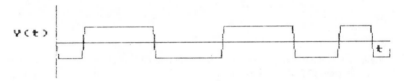

Figure 2.7.   A telegraph signal, in which the times between the changes of
state are Poisson distributed (see §3.2.1). Note that these data and those shown
in figure 2.6 are segments of signals which have *identical* spectral density
functions but, obviously, quite different *amplitude*-domain statistics.

What is immediately obvious from the display, however, is that the
amplitude-domain statistics are completely different from those of the
earlier signal. In the latter case, $y(t)$ was distributed normally (see
§2.5), whereas for the telegraph signal, $y(t)$ can only be either $+A$ or
$-A$ with equal probability. Again, a rearrangement of this signal in
time (to obtain, say, a standard 'square' wave) would completely alter
the spectral content without affecting the amplitude probability distribution in the least.

The reader should endeavour to grasp this distinction between amplitude- and time-domain statistics as soon as possible, since it lies at the heart of the various criteria required for making quantitative measurements (whether analogue or digital) of the signal characteristics actually required.

⟩2.9   Software instructions

From the basic index the user should run the CHAP2 program. The top third of the screen will then contain the material shown at the top of figures 2.1–2.7 (above the solid line). To display any of the signals, simply select the appropriate heading using the cursor keys and then press the RETURN key. Subsequent use of the cursor keys removes the display and allows choice of an alternative option. Some of the options allow further demonstrations, which can be obtained by repeated use of the RETURN key after the initial display has been produced. They are of two types.

Firstly, an infinite variety of three-component Fourier signals can be displayed by continuing the 'almost periodic' option and entering the required amplitudes, frequencies and phases as instructed on the screen. Similarly, a noisy sine wave, with any desired signal-to-noise ratio, can be displayed by continuing the noisy sine-wave option and entering the required value as instructed (see §§2.3 and 2.6, respectively). In both cases, on completion of the display, another screen prompt allows further examples to be generated or the user may return to the basic CHAP2 menu.

Secondly, in the case of the Gaussian noise, correlated noise and the telegraph signals (see §§2.5, 2.7 and 2.8, respectively), continuous oscilloscope-type traces can be obtained by continuing the appropriate option prompt. A 10 000-point data sequence is automatically loaded into the computer memory and plotted on the screen. The signal repeats itself after approximately 45 s, but this is long enough to give the appearance of a continuous signal. A fixed display of any part of the sequence is obtained by using the SPACE bar as instructed, which anyway is required to return the original menu. These data sets have all been generated using standard simulation techniques which will not be discussed here; the methods used are outlined in §4.6.1. These three signals are used for the digital data analysis described in Chapters 4 and 5; so these display options are useful for reminding oneself what the signal looks like, while considering the necessary sampling criteria. The fourth signal used for the data analysis is a set of 19 456 random numbers (as described in §2.4).

The user may return to the main index by using the ESCAPE key.

〉 Chapter 3

〉 Quantitative Description of Signal Content

We consider in this chapter the basic descriptive properties of a stationary signal. These fall naturally into two distinct classes. Firstly, the signal can be described simply in terms of the variation in its amplitude. The basic property from which all amplitude-domain statistics such as the mean value or the variance derive is the probability density function. This is discussed in detail in §3.1, together with the relationships between it and the secondary characteristics (such as the mean value). In many textbooks, definitions of these secondary characteristics are developed first, often without any reference to the probability distribution at all, but we wish here to emphasise that all the former are, in fact, just the various moments of the latter. They can of course be measured without any reference to this fact, but that tends to hide the importance of the *amplitude* bandwidth of the measuring devices.

It is also important to note that, given a particular amplitude probability density function, the signal can be 'arranged' in *time* in an infinite variety of ways. In that sense, the amplitude statistics are independent of the time structure of the signal, as mentioned in Chapter 2. Comparison of figures 2.4 and 2.6 provides an example.

Secondly, the signal can be described in terms of its time-domain behaviour. The most obvious ways are via the autocorrelation function or the frequency spectrum. These are discussed in detail in §3.2 but, again, it should be emphasised that signals with identical spectra can have entirely different amplitude probability density functions. Figures 2.6 and 2.7 provide an example. These remarks do not, of course, imply that the probability density function and, say, the spectral density function are entirely unrelated. The links will be explored in §3.2.

It very often happens that, for a particular physical process leading to a continuous stationary variation of a measured quantity, descriptions of the signal content are much more appropriate in time-domain terms than in terms of amplitude-domain statistics, or vice versa. In the context of signal collection and analysis, however, it is usually helpful to have some idea of the likely qualitative nature of the signal in both frames of reference.

〉3.1   Amplitude-domain statistics

3.1.1 The probability distribution

Consider a stationary signal $x(t)$. At any particular instant, there exists a specific probability that the signal amplitude lies between x and $x + \Delta x$. If T_x is the amount of time, summed throughout the total signal time T, during which $x(t)$ lies in this range, then this probability is just T_x/T. As Δx tends to zero, so must T_x/T (there is an infinitesimally small likelihood that $x(t)$ has a *specific* value, except in the special case of a constant-amplitude signal) but a probability density function $p(x)$ can be defined for small Δx by

$$p(x) = \lim_{\Delta x \to 0} [T_x/(\Delta x T)] \tag{3.1}$$

Unless the signal is band limited (in the amplitude sense), T_x/T must exceed zero but be less than unity whatever particular values of x and Δx are chosen. However, real signals are almost always limited to a certain amplitude range so that, for example, $x(t) \leq r$, say. The probability corresponding to $x > r$ must be zero, and the probability density function $p(x)$ is then not defined for $x > r$. It is important to note that equation (3.1) contains no specific reference to the time-domain structure of the signal. Given an observational period T, $p(x)$ exactly describes the amplitude-domain statistics within that period, but it may have little resemblance to the $p(x)$ that would have been obtained for a different observational period whether of the same duration or not. For example it is straightforward to show that for $x(t) = a \sin(\omega t)$, an observation period T equal to the period $2\pi/\omega$ of the signal gives a probability density function

$$p(x) = \begin{cases} 1/\pi(a^2 - x^2)^{1/2} & |x| \leq a \\ 0 & |x| > a. \end{cases}$$

This is the usual result quoted for the probability density function of a sine wave. However, it is obvious that, if T were a non-integral multiple of $2\pi/\omega$, a very different result would be obtained unless the integral value of $T/(2\pi/\omega)$ were sufficiently large. If $x(t)$ were a truly random signal, one would expect $p(x)$ to tend towards some limiting value (for each x) as T approached infinity. Then equation (3.1) could be written as

$$p(x) = \lim_{T \to \infty} [\lim_{x \to 0} (T_x/\Delta x \, T)] \qquad (3.2)$$

but it is clear that the appropriateness of such a definition depends somewhat on the nature of the signal. We therefore prefer equation (3.1) as the basic definition for $p(x)$ and encourage careful thought when undertaking experimental determinations of it.

To help to fix ideas concerning the probability density function, which is really the fundamental analytic tool for investigating the amplitude statistics of a signal, we now give examples of typical (ideal) distributions, including those corresponding to some of the signals described in Chapter 2. These can be displayed on the screen using the CHAP2 program, as described in §3.4.

3.1.1(a) Constant signal A signal consisting simply of a mean (DC) component of amplitude a, say, will have a probability density function which can be described by a standard delta function

$$p(x) = \begin{cases} \delta(a) & x = a \\ 0 & x \neq a \end{cases}$$

where $\delta(s)$ is defined by

$$\lim_{\varepsilon \to 0} \left(\int_{a-\varepsilon}^{a+\varepsilon} \delta(s) \, ds \right) = 1$$

and $\delta(s) = \infty$ for $s = a$.

3.1.1(b) Telegraph signal The telegraph signal discussed in §2.8, with $x(t) = \pm a$, has a 'double-delta-function' probability density distribution which can be written as

$$p(x) = \begin{cases} \frac{1}{2}\delta(a) & x = |a| \\ 0 & x \neq a. \end{cases}$$

This assumes that the mean value of $x(t)$ is zero. If it is not, then the distribution above is simply shifted along the x axis by the mean value.

3.1.1(c) Periodic signals As stated above, the probability density function for a sinusoidal signal observed over a sufficiently large (or integral) number of cycles is given by

$$p(x) = \begin{cases} 1/\pi(a^2 - x^2)^{1/2} & |x| \leq a \\ 0 & |x| > a \end{cases}$$

where a is the signal amplitude. This is shown in figure 3.1. If the signal contained a mean DC level $(x = a_0 + a \sin t)$, then the $p(x)$ function would again simply be shifted along the x axis by a_0, i.e.

$$p(x) = \begin{cases} 1/\pi[a^2 - (x - a_0)]^{1/2} & |x - a_0| \leq a \\ 0 & |x - a_0| > a. \end{cases}$$

A similar transformation is appropriate in the presence of mean DC components irrespective of the nature of the signal; consequently, we shall assume that $a_0 = 0$ in all the following.

Note the importance of recognising that $p(x)$ is *not* a probability. It is clearly not correct to state that the probability that x takes the value, say, $0.99a$ is $1/\pi(1 - 0.99^2)^{1/2} = 2.26$! It is actually given by

$$\lim_{\varepsilon \to 0} \int_{0.99a-\varepsilon}^{0.99a+\varepsilon} p(x)\, \mathrm{d}x = 0$$

What figure 3.1 does indicate is that the probability that the signal amplitude lies in a range ε near $x = a$ is much higher that it is near, say, $x = 0$. Inspection of the basic signal makes this obvious, of course.

Note also that, because a sine wave is essentially a deterministic signal, a probabilistic description is only really appropriate on the assumption that the initial *phase* (for each sample signal) is a random variable. This is perhaps more evident if a signal composed of two or more sinusoidal components of different amplitudes and frequencies is considered. Unless the different components are uncorrelated (i.e. have random phases), the probability distribution of the total signal must be strongly dependent on the nature of the phase and frequency relationships between the components. In other words, the probability distribution is dependent on the time structure of the signal; in these circumstances, it is not generally very helpful to use a probabilistic description at all. However, if the individual components are uncorrelated, it is possible to deduce a probability distribution which will be quite independent of the phases and frequencies of the individual components.

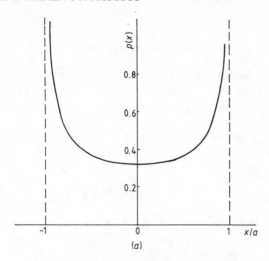

(a)

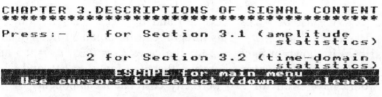

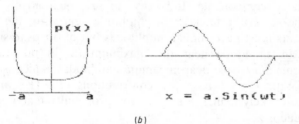

(b)

Figure 3.1. The probability density function of a periodic signal of amplitude $\pm a$.

3.1.1(d) Complex periodic signals As an example, consider the case of a signal $z(t)$ arising from the uncorrelated superposition of two sinusoidal components

$$x(t) = a_1 \sin(\omega_1 t + \theta_1)$$

and

$$y(t) = \sin(\omega_2 t + \theta_2).$$

θ_1 and θ_2 are independent random variables. For, say, $\omega_2 < \omega_1$ and provided that the signal is observed over a sufficiently large number of cycles of ω_1, then

$$p(z) = \begin{cases} \int_{-a_1}^{a_1} p_1(x) p_2(z - x)\, dx & |z| < 1 - a_1 \\ \int_{z-a_1}^{a_1} p_1(x) p_2(z - x)\, dx & 1 - a_1 \leqslant |z| \leqslant 1 + a_1 \\ 0 & |z| > 1 + a_1. \end{cases} \quad (3.3)$$

Using, $p_1(x) = 1/\pi(a_1^2 - x^2)^{1/2}$ and $p_2(y) = 1/\pi(1 - y^2)^{1/2}$, we obtain

$$p(z) = \frac{1}{\pi^2} \int_{\theta_2}^{\theta_1} \frac{d\theta}{[a_1^2 - (z - \cos \theta)^2]^{1/2}}$$

for $|z| < 1 - a_1$, $\theta_1 = \cos^{-1}(z - a_1)$ and $\theta_2 = \cos^{-1}(z + a_1)$ or zero for $1 - a_1 \leqslant |z| \leqslant 1 + a_1$. This integral can be obtained numerically, and $p(z)$ is shown in figure 3.2 for the cases $a_1 = 0.25$, $a_1 = 0.5$ and $a_1 = 1.0$. Note the contrast between the $p(z)$ for $a_1 = 1.0$, and what it would have been for two perfectly correlated components of the same frequency. In the latter case, $z = 2\sin(\omega t + \theta)$ so that $p(z) = 1/\pi(4 - z^2)^{1/2}$, which is very similar to the mirror image (about $z = 0$) of the distribution corresponding to the sum of two uncorrelated components. It is easy to see, qualitatively, why this should be so. For z to have an amplitude near zero, the individual components must have similar amplitudes but of opposite sign whereas, for an amplitude near the maximum (2, in this case), both components must have near-maximum amplitudes, which is much less likely. (With $x = 1 - \varepsilon$ say, y contributions of $-(1 - \varepsilon)$ are very much more likely than $1 + \varepsilon$, since the probability that y is near the latter value is zero!)

It must be emphasised that the distributions shown in figure 3.2 are *not* typical of those corresponding to the complex periodic signals discussed in Chapter 2. These latter signals arise from the sum of strongly correlated components and so have no 'typical' probability density function; each would be highly individual and dependent on the phase and frequency relationships between the separate components. Experimental data are hardly ever of this type; when they are,

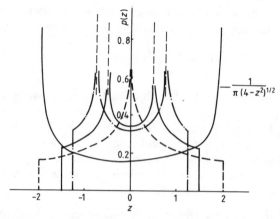

Figure 3.2. The probability density function of the sum of two uncorrelated periodic signals of amplitudes unity and a_1: — · —, $a_1 = 0.25$; ——, $a_1 = 0.5$; — — —, $a_1 = 1.0$.

the time-dependent parameters discussed in §3.2 are generally much more useful tools than the probabilistic description.

The addition of a third uncorrelated sine-wave component of unit amplitude to the example above with $a_1 = 1$ removes the singularity in $p(z)$ at $z = 0$, and not many further components are required to give a $p(z)$ distribution very like the normal distribution (discussed below). In fact, as the number of uncorrelated sine-wave components in the signal increases, the probability distribution function of the signal itself tends towards the normal or Gaussian distribution (see below). This is a direct result of the central-limit theorem, which in its most general form states that the sum of n independently distributed random variables itself tends to a normal distribution as n tends to infinity. This is relatively straightforward to prove if each of the independent variables is normally distributed, but otherwise is true only under certain conditions (Feller 1970).

3.1.1(e) Random noise The case of band-limited random noise is particularly simple. Recall that the signal was defined by

$$y(t) = 2a[0.5 - \text{RND}(1)_t]$$

(Chapter 2.4 and figure 2.3(b). Since $\text{RND}(1)_t$ is (in BBC BASIC) a

function which produces numbers between 0 and 1 with equal probability, the density function is simply

$$p(y) = \begin{cases} 1/2a & |y| \leq a \\ 0 & |y| > a. \end{cases}$$

There are hardly any natural physical processes which give rise to signals of this sort, but the distribution function is nevertheless important. Perhaps the best-known applications of it, in the present context at least, are in the calculation of quantitisation or rounding errors arising from digitisation of a continuous analogue signal or the inevitable truncation in digital arithmetic computations. Chapter 4 gives an example of the former.

3.1.1(f) Gaussian noise Many physical processes lead to amplitude statistics of the variable in question which conform closely to that described by the normal or Gaussian probability density function:

$$p(x) = (1/\sigma\sqrt{2\pi})\exp[-(x - \mu)^2/2\sigma^2]$$

where μ and σ are the mean (DC) value and the standard deviation, respectively.

Since, in addition, this is the distribution to which, according to the central-limit theorem and under certain conditions, the sum of n other distributions of any type will eventually tend as $n \to \infty$, it plays a central role in probability analysis. White noise (defined strictly as a signal having equal power at all frequencies) is usually taken as Gaussian, for that accurately describes the noise occurring in, say, electronic components.

Figure 3.3 shows $p(x)$ for the case of zero mean ($\mu = 0$) and unit variance ($\sigma = 1$), but distributions with any other value can always be made to collapse on this 'standard' distribution by plotting $\sigma p(x)$ against $(x - \mu)/\sigma$. This is, in fact, often done after deduction of the signal mean and variance in order to determine how far from normality the amplitude-domain statistics actually are.

3.1.1(g) Noisy periodic signals Whilst it is very rare that a natural physical process demonstrates purely oscillatory behaviour, it is quite common to find processes which are basically oscillatory but are 'contaminated' by random noise. Even if the basic process *is* purely periodic, its measurement will often involve the addition of random noise because of the transducer characteristics. Sometimes, the signal-to-noise ratio is so low that it is difficult to discern the basic

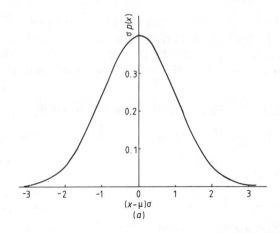

(a)

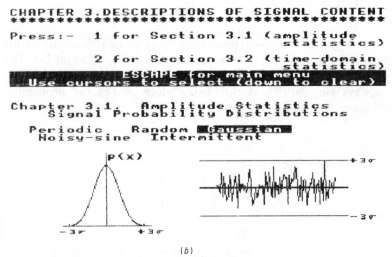

(b)

Figure 3.3. The normal (Gaussian) probability density function.

periodicity. (Such signals can be produced on the computer screen by selecting the appropriate option in CHAP3 program.) There are various techniques for recovering the underlying periodicity, all based on the time-domain characteristics of such signals. Some of these are outlined in Chapter 5. The probability distributions of these types of signal depend on the nature of the noise itself and on the signal-to-noise ratio and can be readily obtained by straightforward application of relationships such as equation (3.3).

If the noise is random but with $p(y) = 1/2r$, such as might be introduced by digitising a perfect sine wave (see §4.1), then the probability distribution of the total signal, consisting of the sum of the underlying periodic component $x = a\sin(\omega t)$ and the random noise, is given by

$$p(z) = \begin{cases} (1/2\pi r)\{\cos^{-1}[(z - r)/a] - \cos^{-1}[(z + r)/a]\} & |z| < a - r \\ (1/2\pi r)\cos^{-1}[(z - r)/a] & a - r \le |z| \le a + r \\ 0 & |z| > a + r \end{cases}$$

if $r < a$, and

$$p(z) = \begin{cases} 1/2r & |z| < r - a \\ (1/2\pi r)\cos^{-1}[(z - r)/a] & r - a \le |z| \le r + a \\ 0 & |z| > a + r \end{cases}$$

if $r > a$. Figure 3.4 shows typical probability density distributions for cases where r/a equals 0 (a normal sine wave), 0.25, 0.5 and 1.0. The singularity in $p(z)$ for $r/a = 0$ is removed as soon as $r \ne 0$, but there are always singularities in $dp(z)/dz$, which makes the probability functions quite distinctive.

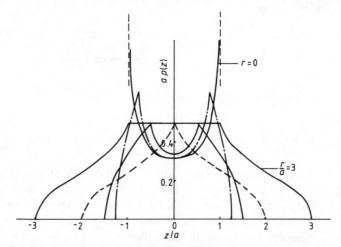

Figure 3.4. Probability distributions of the sum of a period signal of amplitude $\pm a$ and random noise of range $\pm r$: — · —, $r/a = 0.25$; ——, $r/a = 0.5$; — — —, $r/a = 1.0$.

If, as is more likely, the noise is Gaussian, the probability distribution of the total signal is given by

$$p(z) = \frac{1}{\pi\sqrt{2\pi\sigma^2}} \int_{-\pi/2}^{\pi/2} \exp[-(z - a\sin\theta)^2/2\sigma^2]\,d\theta$$

(provided that the noise has zero mean). This is not readily integrable, but can easily be computed numerically for given signal-to-noise ratios. Figure 3.5 shows distributions for cases with σ equal to 0 (the basic sine-wave probability function) 0.25, 0.5 and 1.0. Note particularly that relatively little noise has to be added to the signal before the 'double-peak' characteristic of the $p(z)$ disappears entirely. In such circumstances, it would clearly be impossible to discern any underlying signal periodicity from the shape of the probability distribution, although time-domain signal analysis would clearly reveal it (see §3.2).

3.1.1(h) Intermittent signals When the signal contains two or more components which only occur during different time periods, the probability distribution must clearly have a different form from that if the components occurred simultaneously. Such signals usually occur in practice when the physical processes giving rise to the individual signals switch intermittently from one to another. As a simple example, consider the case of two Gaussian-type processes, with means of μ_1 and μ_2 and standard deviations of σ_1 and σ_2. Assume that the first process occurs during a proportion α of the total observational period, with the second process occurring for the remainder of the period. The probability density function of the signal, measured over the whole period, must then be given by

$$p(z) = (\alpha/\sigma_1\sqrt{2\pi})\exp[-(z - \mu_1)^2/2\sigma_1^2]$$
$$+ [(1 - \alpha)/\sigma_2\sqrt{2\pi}]\exp[-(z - \mu_2)^2/2\sigma_2^2].$$

Figure 3.6(a) shows two cases for which $\mu_1 = 0$, $\sigma_1 = 1$, with $\mu_2 = 0$, $\sigma_2 = 0.25$ and $\mu_2 = 1$, $\sigma_2 = 0.5$. The corresponding signal for the latter case is shown in figure 3.6(b) (screen version). If the two components do not have the same mean values, the resulting distribution is inevitably skewed (asymmetric), whereas greatly unequal standard deviations lead to very peaky distributions. However, it should be emphasised that quite similar probability distributions can be generated by essentially homogeneous (single-component) signals so, as should by now be clear from preceding examples, it is not always possible to make unambiguous deductions about the signal content on the basis of the shape of probability distributions.

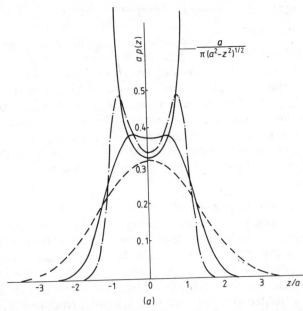

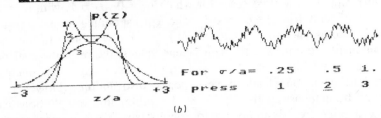

Figure 3.5. Probability distributions of the sum of a periodic signal of amplitude $\pm a$ and white noise of standard deviation σ: — · —, $\sigma/a = 0.25$; ——, $\sigma/a = 0.5$; — — —, $\sigma/a = 1.0$.

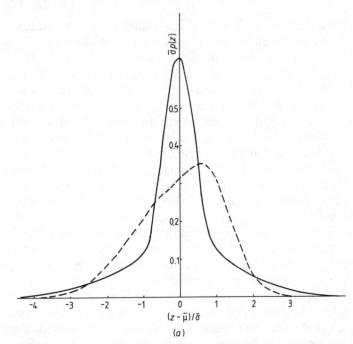

(a)

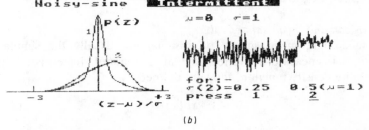

(b)

Figure 3.6. Probability distributions of signals containing two components occurring alternately (the proportional time for the second component is α): ——, $\alpha = 0.5$, $\sigma_2 = 0.25$, $\mu_2 = 0$; ————, $\alpha = 0.25$, $\sigma_2 = 0.5$, $\mu_2 = 1.0$. $\bar{\sigma}$ and $\bar{\mu}$ are the combined standard deviation and mean, given by $\bar{\sigma} \stackrel{!}{=} \sigma_2 + (1 - \sigma_2)\alpha$, $\bar{\mu} \stackrel{!}{=} \mu_2(1 - \alpha)$.

3.1.1(i) Final remarks It should be emphasised that, in all the cases discussed in this section, the time structure of the signal is quite irrelevant as far as the amplitude statistics are concerned. As a 'thought exercise', for example, it is possible to imagine an integral number of cycles of a periodic signal and then to interchange different parts of it along the time axis in a quite arbitrary fashion, so changing its time history catastrophically, but without altering the amplitude-domain statistics in the least. Similarly, one could rearrange a long section of the telegraph signal so that it consisted simply of a constant value a for time $T/2$ followed by a level a for the remaining time $T/2$. This would modify the spectral content of the signal markedly but the probability density function would remain just the same.

In a practical context, very often only the mean value and perhaps the standard deviation of a signal are required, rather than the complete probability density function. However, these are defined analytically as the first and second moments, respectively, of the probability density function (see the following sections) and it is often quite helpful to have some feel for the kinds of signal that produce particular probability distributions. Malfunctions of measuring equipment, for example, can sometimes be detected by inspecting measured probability distributions. Furthermore, the nature of the probability distribution greatly affects some of the sampling criteria that must be applied in order to obtain measurements of a given accuracy; so it is always wise to have some idea about the overall amplitude-domain statistics of the signal, even if that requires (at least rough) measurements of the complete probability density function. As we demonstrate in later chapters, such measurements are often very easy in the case of digitised signals.

3.1.2 Mean and mean-square values
The mean and mean square of a signal are usually the simplest quantities to measure and often all that is required. In terms of the probability density function, they are defined by

$$\bar{x} = \int_{-\infty}^{\infty} x p(x) \, \mathrm{d}x \tag{3.4}$$

and

$$\overline{x^2} = \int_{-\infty}^{\infty} x^2 p(x) \, \mathrm{d}x \tag{3.5}$$

respectively; here and throughout the remainder of this text overbars denote average values. These expressions follow directly from the definition of the probability density function (equation (3.1)) and express the fact that the various mean values are just appropriate weighted averages of the latter. Since $\overline{x^2}$ is the total mean-square value, it is often thought of as the true energy in the signal. Quite frequently, it is helpful to consider only the energy of the fluctuating part of the signal. This is given by

$$\sigma^2 = \overline{x^2} - \bar{x}^2 = \int_{-\infty}^{\infty} (x - \bar{x})^2 p(x)\,dx \qquad (3.6)$$

and is usually termed the variance of the signal. σ is the standard deviation (or more loosely the root mean square (RMS)), which is a simple measure of the width of the probability distribution. Note that σ often has less physical significance than $\overline{x^2}$ or σ^2 and in many contexts it is more useful to consider the latter rather than the 'square root of energy'.

Whilst it is possible to measure mean and mean-square values via the probability density function using equations (3.4) and (3.5), it is of course much more common to obtain them by simple averaging in time, using

$$\bar{x} = \lim_{T \to \infty} \left(\frac{1}{T} \int_0^T x\,dt \right) \qquad (3.7)$$

and

$$\overline{x^2} = \lim_{T \to \infty} \left(\frac{1}{T} \int_0^T x^2\,dt \right). \qquad (3.8)$$

Note that, as in the case of $p(x)$ (see §3.1.1), the earlier definitions of $\bar{x}$ and $\overline{x^2}$ contain no specific reference to the time structure of the signal. Given $p(x)$ measured over time T using equation (3.1) the resulting $\bar{x}$ and $\overline{x^2}$ will be the values appropriate for a signal sample of length T and these may or may not be close to those that would have been obtained for a different observational period. The two sets of definitions are only identical when $p(x)$ is considered to be defined by equation (3.2).

Most of the signals described in the previous sections had zero mean value for convenience, but their variances were never zero— only a signal with constant amplitude for all time can have $\sigma^2 = 0$. The telegraph signal $x(t) = \pm a$ has a variance of a^2; the band-limited

random noise with $p(x) = 1/2a$ for $-a < x \leq a$ has $\sigma^2 = a^2/3$ and
the Gaussian noise signal with $p(x) = (1/\sigma\sqrt{2\pi})\exp[-(x-\mu)^2/2\sigma^2]$
has $\bar{x} = \mu$ and a variance of σ^2. It is left as a simple exercise for the
reader to deduce these results from equations (3.4) and (3.6). Note
that in the last case the results indicate *why* $p(x)$ for Gaussian
variables is defined as it is; the probability density function of all
Gaussian variables can be collapsed onto a single normalised curve by
plotting $\sigma p(x)$ against $(x - \mu)/\sigma$. Hence we have the common term
normal distribution for the above $p(x)$ distribution.

3.1.3 Higher-order moments

The mean and mean-square values defined by equations (3.4) and
(3.5) are simply the lowest two integral 'moments' of the probability
density function. In general, the nth moment is defined by

$$\overline{x^n} = \int_{-\infty}^{\infty} x^n p(x)\,\mathrm{d}x$$

and higher-order moments $(n > 2)$ are sometimes used to deduce
certain physical characteristics of the signal. For example, any signal
in which extreme low values are more likely than extreme high values
(or vice versa) will have an asymmetric probability distribution
function and consequently non-zero values of all the odd-order
moments $(n = 3, 5, \ldots)$. Such distributions are usually called
'skewed'; there is an example in figure 3.6 where the skewness arises
from the fact that the component 'intermittent' signals have different
mean values. The other example in figure 3.6 is of a 'peaky' signal,
which is symmetric and therefore has zero odd-order moments, but
the even-order moments are quite different from the Gaussian values
(see below).

In comparing higher-order moments of different signals, it is
common to normalise them by the appropriate power of the standard
deviation. Thus, for the general case in which $\bar{x} \neq 0$, we define

$$\gamma_n = \frac{\overline{x^n}}{\sigma^n} = \int_{-\infty}^{\infty}(x-\bar{x})^n p(x)\,\mathrm{d}x \bigg/ \left(\int_{-\infty}^{\infty}(x-\bar{x})^2 p(x)\,\mathrm{d}x\right)^{n/2} \qquad n \geq 2.$$

$$(3.9)$$

For $n = 2$, $\gamma_n = 1$ but, for all higher-order even moments, $\gamma_n > 1$. It
is easy to demonstrate the following results, for even-order moments.

(1) Telegraph signal: $p(x) = \frac{1}{2}\delta(a)$, $x = \pm a$; $\gamma_n = 1$.
(2) Random noise: $p(x) = 1/2a$, $-a < x < a$; $\gamma_n = 3^{n/2}/(n+1)$

(3) Gaussian noise: $p(x) = (1/\sigma\sqrt{2\pi})\exp(-x^2/2\sigma^2)$;
$\gamma_n = 2^{n/2}\Gamma((n+1)/2)/\sqrt{\pi} = 1 \times 3 \times 5 \times \ldots \times (2k - 1)$, $n = 2k$.

Note how rapidly the γ_n rises with n for the noise signals. In the case of Gaussian noise, $\gamma_4 = 3$ (where γ_4 is normally termed the flatness factor) whereas $\gamma_8 = 105$. The contribution to γ_n from the 'tails' of the probability density function becomes increasingly important as n rises and this has significant implications for the measurement of such quantities (see Chapter 4.2).

As in the case of the mean and mean-square values, the most straightforward way to obtain the higher-order moments is usually via simple time integration, using

$$\overline{x^n} = \lim_{T\to\infty}\left(\frac{1}{T}\int_0^T (x - \bar{x})^n \, dt\right). \tag{3.10}$$

Unless the amplitude ranges of the measuring devices, whether analogue or digital, are sufficiently wide, errors in measurement of high-order moments can be significant. It is often possible to calculate the errors that would arise in particular cases. This is discussed at greater length in Chapter 4, but it is salutary to recognise that, if the integration bandwidth in equation (3.9) is limited to, say, plus and minus three standard deviations, the resulting values of γ_4, γ_6 and γ_8 for a Gaussian signal are about 9%, 16% and 21% too low, respectively, though $p(x)$ at $(x - \bar{x})/\sigma = 3$ has fallen to 1% of its maximum value!

It is, incidentally, considerations of this sort which emphasise the difficulty of making extreme-value predictions on the basis of small numbers of samples, a problem that bedevils the statistical aspects of the economic, social and medical sciences in particular.

〉3.2 Time-domain statistics

3.2.1 The autocorrelation function
Most time-dependent signals arising from specific physical processes have some structure in time; that is, the signal amplitude at one time is not independent of its value at preceding times. This is obviously the case for deterministic signals such as a simple sine wave, but it is also often true for random data. In Chapter 2, typical Gaussian signals both with and without a time structure were presented (see

figures 2.4 and 2.6). A simple measure of the degree of dependence of the signal amplitude at one time on amplitudes at other times is provided by the autocovariance, defined by

$$C(\tau) = \lim_{T \to \infty} \left(\frac{1}{T} \int_0^T x(t)x(t + \tau)\,\mathrm{d}t \right) \tag{3.11}$$

where $x(t)$ is the signal amplitude at time t and $x(t + \tau)$ is the value at a time τ later. τ is called the lag time. $C(\tau)$ is simply the time averaged product of the two values.

It is clear that $C(0) = \overline{x^2}$, the mean square of the signal. If $\bar{x} \neq 0$, it is more common to use the autocorrelation function

$$R(\tau) = \lim_{T \to \infty} \left(\frac{1}{T} \int_0^T [x(t) - \bar{x}][x(t + \tau) - \bar{x}]\,\mathrm{d}t \right) \tag{3.12}$$

so that $R(0) = \sigma^2$, the signal variance. For random signals, if the time lag is long enough, there is no correlation (on average) between the two (AC) values of the signal; one could say that the signal has a 'finite memory'. In that case it follows from the above definitions that $C(\infty) = \bar{x}^2$ (since $R(\infty) = C(\infty) - \bar{x}^2$). So the mean value is given by $\bar{x} = \sqrt{C(\infty)}$. This is not true for deterministic signals such as sine waves, since these have a 'perfect memory'; it is always possible to predict future values exactly on the basis of current or past values. For convenience, in this section, we shall assume that $\bar{x} = 0$ in all examples so that the autocovariance and the autocorrelation functions are identical.

It is also possible to express the autocorrelation function in terms of probability functions. The probability that $x(t)$ lies in the range $x_1 < x(t) < x_1 + \Delta x_1$ while, simultaneously, $x(t + \tau)$ lies in the range $x_2 < x(t + \tau) < x_2 + \Delta x_2$ is given by

$$\lim_{T \to \infty} [T(x, y)/T]$$

where $T(x, y)$ is the total amount of time that $x(t)$ and $x(t + \tau)$ lie in these ranges simultaneously during the observation time T. Then the joint probability density function $p(x_1, x_2)$ is defined by

$$p(x_1, x_2) = \lim_{\substack{\Delta x_1 \to 0 \\ \Delta x_2 \to 0}} \lim_{T \to \infty} \{(1/T)[T(x, y)/(\Delta x_1 \Delta x_2)]\}.$$

This is an obvious extension of the $p(x)$ definition discussed in §3.1 Then if we denote $x(t)$ by x_1 and $x(t + \tau)$ by x_2, the autocovariance must be given by

$$C(\tau) = \int_{-\infty}^{\infty} \int_{-\infty}^{\infty} x_1 x_2 p(x_1, x_2) \, \mathrm{d}x_1 \, \mathrm{d}x_2. \qquad (3.13)$$

If the average value of the signal is zero, this is also $R(\tau)$. It must be emphasised that, whilst, for a sufficiently long observation time, $p(x_1)$ and $p(x_2)$ will not depend in any way on the time structure of the signal, $p(x_1, x_2)$ certainly will. (For stationary random signals $p(x_1)$ will normally be identical with $p(x_2)$.)

In some standard texts the autocorrelation function is defined in non-dimensional terms, by dividing the $R(\tau)$ above by the signal variance. This makes the autocorrelation unity for $\tau = 0$. We shall use $R(\tau)$ in both ways in the following; it will always be evident from the context whether a normalised function or a function which has not been normalised is meant. In either case, for stationary signals, $R(\tau) = R(-\tau)$.

It should be noted that the prefix 'auto' is used because these functions describe the time structure of a *single* signal, rather than relationships between two or more signals (at the same or different times). The latter are described by, for example, *cross* correlations. Multi-signal statistics are not covered in the present text, but the essential ideas are quite similar.

We now give some examples of autocorrelation functions, using some of the typical signals previously described.

3.2.1(a) Periodic signals For a signal $x(t) = a \sin(\omega t + \phi)$, it is easy to show that the autocorrelation is given by $(a^2/2) \cos(\omega \tau)$. This is illustrated in figure 3.7. It is important to note that all phase information in the original signal is lost and that the period of the autocorrelation function is the same as that of the original signal. Note also that $R(0) = a^2/2$, the mean square of the signal, as expected.

3.2.1(b) Uncorrelated noise 'White' noise is usually defined as a Gaussian process in which the signal amplitude at any particular time has, on average, no dependence on amplitudes at other times. The autocorrelation function is then simply $R(\tau) = \delta(\tau)$, where $\delta(\tau)$ is the Dirac delta function (defined in §3.1.1(a)—it is infinite for $\tau = 0$ but zero otherwise) (figure 3.8). The signal has constant energy at all frequencies (see §3.2.2), so perfect white noise can never exist in practice since it would have infinite power. White noise is always band limited, in the frequency sense, and it will be shown later that, if the bandwidth is B (i.e. no energy at frequencies higher than B)

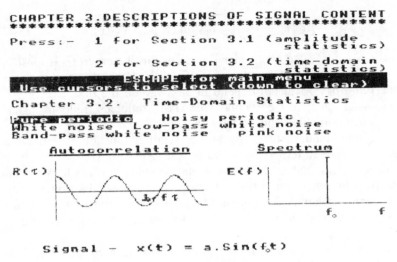

Figure 3.7.   Autocorrelation and spectrum of a sine wave.

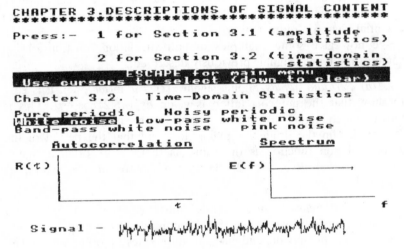

Figure 3.8.   Autocorrelation and spectrum of white noise.

and the variance is σ^2, then the autocorrelation function is given by

$$R(\tau) = \sigma^2\{[\sin(2\pi B\tau)]/2\pi B\tau\}.$$

This is shown in figure 3.9; it is clear that, as $B \to \infty$, $R(\tau) \to 0$ except at $\tau = 0$, when $R(\tau) = \sigma^2$ always.

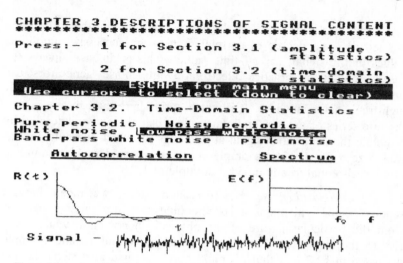

Figure 3.9. Autocorrelation and spectrum of low-pass filtered white noise, assuming a sharp cut-off at frequency f_0.

If the signal is band limited at the lower frequencies as well as the highest, so that it has constant power in a frequency band of width B centred at f_0 ($<B/2$), the autocorrelation function has the form

$$R(\tau) = \sigma^2\{[\sin(\pi B\tau)]/\pi B\tau\} \cos(2\pi f_0\tau).$$

An example is shown in figure 3.10.

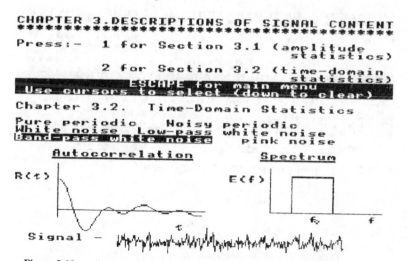

Figure 3.10. Autocorrelation and spectrum of band-pass filtered white noise.

Clearly, in both these cases the act of band limiting the white noise has led to some degree of correlation between successive amplitude values of the signal, for *all* time values. This is, in fact, always the case although in practice electronic white noise generators can produce quite adequate 'quasi'-white noise by having a sufficiently high-frequency bandwidth. It is worth noting here that, in any case, a discrete series of uncorrelated data values can always be generated by sampling the continuous signal at time lags for which the autocorrelation is zero. In the first example above, this would require sampling at intervals equal to any integral multiple of $1/2B$.

3.2.1(c) Correlated noise By correlated noise, we mean here a random signal which has a specific time structure, arising perhaps from the particular nature of the physical phenomenon which gives rise to the signal, so that $R(\tau)$ is non-zero for all τ. The example discussed in §2.7 was that of 'pink' noise, defined as a signal having an exponentially decaying autocorrelation function: $R(\tau) = \exp(-\lambda\tau)$. λ determines the decay rate. Very many natural phenomena lead to signals having a behaviour similar to this, which makes pink noise a particularly relevant analytic generalisation for studying, for example, probable measurement errors in real situations. Figure 3.11 shows some typical pink noise with its corresponding autocorrelation function.

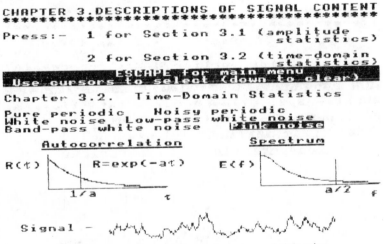

Figure 3.11. Autocorrelation and spectrum of pink noise.

As indicated earlier, signals with identical autocorrelation functions can have quite dissimilar probability density functions. The telegraph signal discussed in Chapter 2 provided a good example of this. It has a probability density function defined by $p(x) = 0.5\delta(a)$ for $x = \pm a$ and $p(x) = 0$ otherwise, independent of the distribution in time of the state changes between $\pm a$. However, if these state changes are arranged according to a Poisson distribution, so that the probability that n changes occur in a time interval τ is given by

$$\Pr(n) = (\lambda\tau)^n \exp(-\lambda\tau)/n!$$

where λ defines the expected number of changes ($E(n) = \lambda\tau$), then it is straightforward to show that the autocorrelation function is given by $R(\tau) = \exp(-2\lambda\tau)$ (for a simple proof, see Bendatt and Piersol (1966)). Consequently, on this basis or on the basis of the signal's energy spectrum (see §3.2.2), the signal is indistinguishable from pink noise; however, it is clearly not a 'noise' signal at all in the usual sense.

3.2.1(d) Noisy periodic signals   Signals which consist essentially of an underlying periodic component with the addition of random noise are fairly common. If the noise has sufficient amplitude, it can be impossible to discern the basic periodicity either by direct inspection of the signal or by looking at its probability density funtion (see §3.1). By contrast, the signal's autocorrelation function will clearly reveal the periodicity, particularly if the noise is uncorrelated (white). The autocorrelation function then consists of the usual cosine function but with a delta function 'spike' at zero lag, as shown in figure 3.12. Defining the signal by

$$x(t) = a\sin(\omega t) + n(t)$$

where $n(t)$ is random noise with a standard deviation of σ, the ratio of the standard deviation of the periodic component to that of the noise is $a/\sigma\sqrt{2}$. In the example shown, this ratio is unity so that $C(0) = a^2/2 + \sigma^2$ and $C(\tau) = (a^2/2)\cos(\omega\tau)$ ($\tau > 0$). Note that we have assumed the noise to be band limited so that it has a finite variance and have ignored the consequent effects on the combined $R(\tau)$.

Quite frequently, the 'noise' arises from a phenomenon having some specific time structure (i.e. it is correlated noise). In that case the 'spike' at zero lag would not exist but the first few cycles in $R(\tau)$ would decay until a time lag beyond which the noise has no significant autocorrelation of its own. Thereafter $R(\tau)$ would be a perfect cosine wave. Another common variant is the case of a phenomenon in which the basic periodic component is modulated by occasional random shifts in phase, in addition to the superimposed noise. After a sufficiently long lag time the autocorrelation would then be expected to be zero and a typical autocorrelation function might have the form $R(\tau) = \cos(\omega\tau)\exp(-a\tau)$.

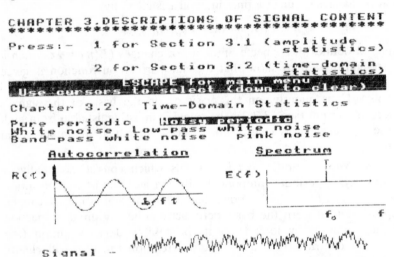

Figure 3.12. Autocorrelation and spectrum of a sinusoidal signal contaminated with white noise.

3.2.1(e) General autoregressive process Some of the signals discussed in the last two sections are, in fact, just fairly common examples of general autoregressive processes. These are usually defined as the output $Y(t)$ from a linear filter whose input $X(t)$ is white noise and for which the input and output are related by

$$a_m(d^m Y/dt^m) + a_{m-1}(d^{m-1}Y/dt^{m-1}) + \ldots + a_0 Y(t) = X(t).$$
$$(3.14)$$

For $Y(t)$ to be stationary, it is necessary that the roots of the characteristic equation

$$a_m p^m + a_{m-1}p^{m-1} + \ldots + a_0 p = 0$$

have negative real parts. It is shown in standard texts (see, e.g., Priestly 1981) that the autocorrelation of $Y(t)$ must satisfy

$a_m(\mathrm{d}^m R(\tau)/\mathrm{d}\tau^m) + a_{m-1}(\mathrm{d}^{m-1} R(\tau)/\mathrm{d}\tau^{m-1}) + \ldots + a_0 R(\tau) = 0\,\tau > 0.$

This has the general solution

$$R(\tau) = A_1 \exp(-\lambda_1\tau) + A_2 \exp(-\lambda_2\tau) + \ldots + A_m \exp(-\lambda_m\tau)$$

where the λ_i are the roots of the characteristic equation given above.

Pink noise can therefore by thought of as the output of a first-order ($m = 1$) linear system so that, in this notation, $R(\tau) = \exp(-a_0\tau/a_1)$. If complex roots occur (for $m > 1$), then the autocorrelation contains terms of the form $\exp(-k_1\tau)\cos(k_2\tau + \phi)$; a signal which has a decaying autocorrelation function could therefore arise from the output of a second-order ($m = 2$) linear system whose input is white noise. Note, however, that, if $\phi = 0$ (so that the autocorrelation is the same as that used in the examples in §3.1), then the process must be a mixed autoregressive and moving average one, for which the right-hand side of equation (3.14) is replaced by

$$\int_0^{t_m} b(u)X(t - u)\,\mathrm{d}u$$

where $b(u)$ is some weighting function which is zero for $t > t_m$.

Signals produced this way would necessarily have Gaussian amplitude-domain statistics (since the linearity of the system precludes any change in the signal's probability distribution). It must therefore be emphasised again that a signal having this particular $R(\tau)$ *could* arise from a quite different physical process and have a non-Gaussian probability density distribution.

3.2.2 The power spectral density function

An alternative but complementary way of describing the time-domain nature of a signal is by way of its energy spectrum. Physically, this is simply a measure of how much energy is contained within the signal in each frequency band. If $x(t, f, \Delta f)$ is that part of the signal $x(t)$ which lies in the frequency band Δf centred at frequency f, then the mean-square value within this band is

$$S^2(f) = \lim_{T\to\infty}\left(\frac{1}{T}\int_0^T x^2(t, f, \Delta f)\,\mathrm{d}t\right). \qquad (3.15)$$

Then the power spectral density can be defined by

$$E'(f) = \lim_{\Delta f\to 0}[S^2(f)/(\Delta f\,\overline{x^2})] \qquad (3.16)$$

where, as usual, $\overline{x^2}$ is the total mean square (or variance σ^2). Note the correspondence between this definition and that of the probability density function (equation (3.1)). In both cases a suitable normalisation is used such that the integral of the distribution over the whole parameter range is unity. In this case,

$$\int_0^\infty E'(f)\,df = 1.$$

An alternative definition for a power spectral density which has not been normalised is sometimes used. Typically,

$$E(f) = \lim_{\Delta f \to 0}\left[\lim_{T \to \infty}\left(\frac{1}{\Delta f\,T}\int_0^T x^2(t, f, \Delta f)\,dt\right)\right]$$

so that $\int_0^\infty E(f)\,df = \overline{x^2}$, the total signal energy. Note that the mean value $\bar{x}$ is just the square root of the DC energy $E(0)$.

It should be clear immediately that there must be at least some relationship between the autocorrelation function and the spectral density function, since $R(\tau) = \overline{x^2}$ when $\tau = 0$ (see §3.2.1); so $R(0) = \int_0^\infty E(f)\,df$. In fact, the two functions form an exact Fourier transform pair:

$$E(f) = 2\int_{-\infty}^\infty R(\tau)\exp(-2\pi i f\tau)\,d\tau$$

$$= 4\int_0^\infty R(\tau)\cos(2\pi f\tau)\,d\tau \qquad (3.17a)$$

$$R(\tau) = \frac{1}{2}\int_{-\infty}^\infty E(f)\exp(2\pi i f\tau)\,df$$

$$= \int_0^\infty E(f)\cos(2\pi f\tau)\,df \qquad (3.17b)$$

The second of each pair of equalities exist because, for real data, $R(\tau)$ is an even function of τ. These relationships hold only provided that the original signal is stationary. They are collectively known as the Weiner–Khintchine therorem and standard texts discuss their proof in detail (see, e.g., Bartlett 1955). Basic definitions and discussions of Fourier transform theory are also available in many specialist texts (see, e.g., Lighthill 1962, Priestly 1981). It should be noted that such texts usually embody many minor differences in definition (and therefore in detailed analysis); we have chosen here those which seem most appropriate for real data.

It is also possible to express the spectral energy density function in terms of the original signal itself. If it is assumed that the signal $x(t)$

has a convergent Fourier integral $C(f)$ given by

$$C(f) = \int_0^\infty x(t)\exp(-2\pi ift)\,dt \qquad (3.18)$$

which for random signals requires ensuring that $x(t) = 0$ outside some finite time range, then it can be shown that

$$E(f) = \lim_{T\to\infty}[|C(f)|^2/T]. \qquad (3.19)$$

$|C(f)|^2$ is the sum of the squares of the amplitudes of the real and imaginary parts of the Fourier integral (i.e. the modulus, in the usual sense). Strictly, equation (3.19) is only true for random signals if the numerator is $\text{ave}_k[|C(f)|^2]$, i.e. an average of $|C(f)|^2$ over k independent ensembles, with $k \to \infty$. This restriction is not necessary for deterministic (periodic) signals but in the context of much experimental signal analysis, in which strong random elements are often present, it has serious implications for the determination of the energy spectrum. Further discussion is deferred to Chapter 5 and, again, standard texts should be consulted for more detailed analysis.

It should be emphasised that the autocorrelation function $R(\tau)$ and spectral density function $E(f)$ are entirely equivalent. There is no information contained in the one which is not present in the other. The difference lies simply in the way in which the information is presented; $R(\tau)$ gives essentially time-domain data whereas $E(f)$ gives the same information but in the frequency domain. In some contexts the former is more physically useful than the latter, and vice versa.

The reader will have already noted in figures 3.7–3.12 the energy spectrum functions corresponding to the autocorrelation function examples discussed in §3.1. We give below the analytic expressions for these spectra and leave it as an exercise for the reader to derive them by performing the appropriate cosine transforms, given by equation (3.17b). For completeness, we include the autocorrelation functions below.

3.2.2(a) Periodic signal (figure 3.7)

$$x(t) = a\sin(2\pi f_0 t + \phi)$$
$$E(f) = (a^2/2)\delta(f - f_0)$$
$$R(\tau) = (a^2/2)\cos(2\pi f_0\tau).$$

3.2.2(b) Uncorrelated noise (figure 3.8) Gaussian amplitude-domain data are used with

$$E(f) = a \qquad f > 0$$

$$R(\tau) = a\delta(\tau).$$

Note here, firstly, that this is not physically very realistic since $\overline{x^2} = \int_0^\infty E(f)\,df = \infty$ and, secondly, that random noise with different amplitude-domain statistics could have identical $E(f)$ and $R(\tau)$.

3.2.2(c) Constant DC signal It is worth noting that in time and frequency domains a constant signal of amplitude $\sqrt{a}$ is the exact inverse of white noise, i.e.

$$x(t) = \sqrt{a}$$

$$E(f) = a\delta(f)$$

$$R(\tau) = a.$$

3.2.2(d) Low-pass white noise (figure 3.9).

$$E(f) = a \qquad 0 \leq f \leq B$$

$$R(\tau) = aB\{[\sin(2\pi B\tau)]/2\pi B\tau\}.$$

3.2.2(e) Band-pass white noise (figure 3.10)

$$E(f) = a \qquad (f_0 - B/2) \leq f \leq (f_0 + B/2)$$

$$R(\tau) = aB\cos(2\pi f_0\tau)\{[\sin(\pi B\tau)]/\pi B\tau\}.$$

Note that in both cases the signal variance is given by $\overline{x^2} = \sigma^2 = aB$.

3.2.2(f) Correlated noise (figure 3.11) Gaussian amplitude-domain statistics are used with

$$E(f) = 4T\sigma^2/(1 + 4\pi^2 f^2 T^2)$$

$$R(\tau) = \sigma^2 \exp(-\tau/T)$$

As stated earlier, this defines pink noise. The telegraph signal described in §§2.8 and 3.1.1 has the same forms of $E(f)$ and $R(\tau)$.

3.2.2(g) Noisy periodic signal (figure 3.12) Provided that the noise is not correlated with the periodic content of the signal, $E(f)$ and $R(\tau)$ are simple additions of those corresponding to the separate components. For a periodic component of $a_1\sin(2\pi ft)$ superimposed on white noise whose energy density is a_2 at all frequencies,

$$E(f) = (a_1^2/2)\delta(f - f_0) + a_2$$
$$R(\tau) = (a_1^2/2)\cos(2\pi f\tau) + a_2\delta(\tau).$$

Again, this is not physically very realistic; the reader will be able to write down other more representative examples.

⟩3.3   Further statistics

The quantities discussed in §§3.1 and 3.2 are not, of course, the only ones that can be used to describe a signal, although they are certainly the most basic. In the natural and engineering sciences there are often requirements to analyse particular portions of a signal which represent isolated, but perhaps repeated, events. As a simple example, consider the electrical signal produced by cardiographic skin sensors. These measure electrical activity at the surface which corresponds to events in the cardiac cycle. The signal is characterised by repeated sequences of the portion shown in figure 3.13, which represents such a cycle. Each sequence is of course different from preceding ones, because of the many and sometimes random physical processes which affect the cardiac cycle. Rather than studying a complete spectral analysis obtained over many cycles, it is often more useful for diagnostic purposes to examine the form of a particular part of the cycle. Both the mean 'shape' and the nature of the variations about this mean may be of interest. Obtaining data of this type requires use of the technique generally known as conditional sampling. This usually involves obtaining averaged data from one part of a signal by using another part as a conditional trigger. Referring to figure 3.13, for example, one might be interested in the shape of the QRS part of the signal in cases where the R peak is abnormally low. Equally, it may be useful to know how the frequency content of the rest of the signal differs from average when the initial R spike is abnormally distorted.

Simple extensions of the definitions of the amplitude- and/or time-domain statistics given earlier can easily be written down for such cases and suitable algorithms derived for measuring them. Obviously these kinds of measurement are very much simpler using digital techniques. The basic ideas are in principle no different from those outlined already (and in Chapter 4). Since this is an introductory text, these other kinds of signal statistics will not be discussed

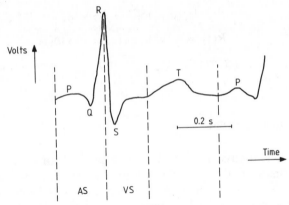

Figure 3.13. A typical electrocardiogram: AS, atrial contraction; VS, ventricular contraction.

further. Despite a rapid increase in the use of conditional sampling in many different fields, it remains true that the more basic statistical functions described in the earlier sections are by far the most commonly used.

⟩3.4 Software instructions

From the basic index the user should run the CHAP3 program. The top part of the screen will contain the material shown at the top of the screen version of figure 3.1 (above the solid line). Using the numeric keys 1 or 2, the amplitude- or time-domain sections, respectively, can be called. To display the required signal, together with the corresponding probability density function or autocorrelation and spectrum functions, the cursor and RETURN keys should be used as before. Some of the amplitude-domain statistics options allow further demonstrations; these are described below.

The noisy sine option shows three probability density functions corresponding to the three possible signals on the right-hand side of the screen. The signal initially displayed (1) has a noise component whose RMS value is one quarter of the amplitude of the periodic component. Signals with two or four times this σ/a ratio can be displayed by using the numeric keys 2 or 3 respectively. Figure 3.5 (screen version) includes the latter example. The three corresponding $p(x)$ distributions are labelled.

Similarly, for the intermittent signal option, two distributions are shown and either of the signals can be displayed as above.

The user can switch between §§3.1 and 3.2 by first clearing the display using the down cursor key, and then using the required numeric key. Pressing the ESCAPE key returns the system to the main index.

〉 Chapter 4

〉 Digital Sampling Criteria: Amplitude-domain Statistics

〉4.1 Introduction

If a signal is to be analysed using analogue electronics only, a major question facing the experimenter—in addition to those concerning instrument amplitude and frequency bandwidth—is that of the required integration time. In most circumstances, this must be long enough to average the lowest frequency content of the signal adequately. In the case of analysis of a digitised signal, the questions become as follows.

(1) How many individual digitised data samples are required?
(2) How rapidly must the digitising be carried out?
(3) What digitiser ·esolution is required?

The last of these questions is a direct result of replacing an analogue signal which has an infinite number of amplitude values by a digital signal which can take only a finite number of values (the amplitude discretisation). This leads to inevitable inaccuracies, generally known as 'quantisation' errors, but these can usually be made insignificant. A discussion is given below (see §4.2). The first two questions are a result of the time-domain discretisation and effectively replace the 'analogue' question concerning the integration time. It should be obvious immediately that they are not entirely independent. It will still usually be necessary to process a sufficiently long period of the signal to allow adequate averaging of any low-frequency content, but clearly this can be done in a variety of ways—the product of the sampling rate and the number of samples gives the sampling period.

48

It should also be obvious that the answers to these two questions will depend to a greater or lesser extent on the nature of the signal itself and on the particular measurement envisaged. For example, if the requirement is to measure the mean square or indeed any characteristic of the amplitude statistics of a white noise signal, it does not matter how rapidly or how slowly the signal is digitised. The measurement accuracy will depend solely on the number of samples used (neglecting quantisation errors). In contrast in the case of a signal for which the autocorrelation function is non-zero at finite lag times, the sampling rate becomes more relevant. If only amplitude-domain statistics were required, it would clearly be foolish to sample at 1 MHz a signal containing a strong periodic component at 1 Hz.

What would normally be required in such a case is a sampling *time* long enough to average this low-frequency content adequately, with a sampling *rate* chosen to give a number of samples large enough to minimise statistical errors arising from the finite number of data points. Higher sampling rates could lead to lower statistical errors (for the same sampling period) but would be unduly wasteful. In the context of time-domain measurements (of spectral and autocorrelation functions), sampling rates are much more crucial, since they determine the highest frequency that can be resolved. It is clearly impossible, for example, to obtain the spectral content of a signal at frequencies of 1 MHz if it is sampled at only 1 Hz.

In this and Chapter 5, these various matters are discussed in some detail. The intention is not to provide a full mathematical treatment of the subject but rather to give the reader help in developing a 'feel' for the problems, whilst providing some basic quantitative guidelines. In §4.2, amplitude quantisation errors are discussed and, in the remaining sections, we concentrate on the errors which arise in amplitude-domain measurements owing to finite numbers of samples. Time-domain measurements are discussed in Chapter 5.

⟩4.2   Quantisation and ranging errors

There are in general two ways of representing analogue data digitally. These are usually referred to as delta modulation and pulse-code modulation. The former uses a one-digit code only and it is effectively the time derivative of the signal amplitude which is transmitted. It is most often used for data transmission over long distances or for speech transmission. For general-purpose signal analysis as in the present context, pulse-code modulation is invariably used so discussion here is restricted to this case, in which the quantisation errors arise from transforming an analogue signal to a digital signal having a number of distinct levels. These amplitude levels (or quanta) are uniquely specified by the n-digit code; for example, in a binary system, 256 amplitude levels implies use of an 8-bit code since $2^8 = 256$. Obviously, as n increases, the quantisation errors must decrease.

Consider the instantaneous amplitude of the analogue signal to be x with the nearest quantising level x_i. Then the error is $x - x_i$ with a mean square error of $\overline{(x - x_i)^2}$. If the incremental quantisation bandwidth is Δx_i then all signal amplitudes within the range $x_i - \Delta x_i/2$ to $x_i + \Delta x_i/2$ will be referred to the ith level in the analogue-to-digital conversion. The mean-square error associated with this level is therefore

$$\sigma_i^2 = \int_{x_i - \Delta x_i/2}^{x_i + \Delta x_i/2} (x - x_i)^2 p(x)\,\mathrm{d}x$$

where $p(x)$ is the probability density function of the analogue signal.

Higher-order moment errors can be similarly expressed. Now, provided that the step size Δx_i is small compared with the total signal amplitude range, the signal can be assumed to be uniformly distributed over Δx_i regardless of its statistical distribution over the whole range. Hence

$$\sigma_i^2 = \Delta x_i^3 p(x_i)/12$$

so that the total mean-square noise can be written

$$\sigma_n^2 = \sum_{i=1}^{N} \frac{\Delta x_i^2\, p(x_i)\, \Delta x_i}{12}.$$

For the simple case of a linear quantisation (all Δx_i steps are of equal size),

$$\sigma_n^2 = \tfrac{1}{12} \Delta x^2 \sum_{i=1}^{N} p(x)\, \Delta x$$

and, since $\Sigma_{i=1}^{N} p(x)\, \Delta x = 1$,

$$\sigma_n^2 = \Delta x^2/12. \tag{4.1}$$

As an example, consider first a random (Gaussian) signal of zero mean and standard deviation σ, quantised into n levels covering an amplitude range of $\pm 3\sigma$. The digital step size would then be $6\sigma/(n-1)$ so that the RMS signal to (quantising) noise error would be

$$\sigma/\sigma_n = \sigma/(\Delta x/\sqrt{12}) = (n-1)/\sqrt{3}.$$

In the case of a simple sine wave of peak-to-peak amplitude $2a$ just filling an n-bit digitiser, the step size would be $2a/(n-1)$ so that the signal-to-noise ratio would be

$$(a/\sqrt{2})[(n-1)/2a]\sqrt{12} = (n-1)\sqrt{\tfrac{3}{2}}.$$

These results are compared in table 4.1 for a range of n.

Table 4.1. Quantisation noise for signals with various $p(x)$ and various quantising levels.

Quantising level n	Number of bits	RMS signal-to-noise ratio		
		Gaussian $p(x)$	Sine wave	Uniform $p(x)$
8	3	4.04	8.6	7
16	4	8.7	18.4	15
32	5	17.9	38.0	31
64	6	36.4	77.2	63
128	7	73.3	155	127
256	8	147	311	255

Note first that the digitisation of the periodic signal is inherently more accurate than that of the Gaussian signal, for a given number of bits. This is a direct result of the very different form of the probability distributions of the two signals. Secondly, errors caused by restricting the digitised amplitude range of the Gaussian signal to $\pm 3\sigma$ and therefore 'clipping' the extreme amplitudes have been ignored. They would clearly be reduced by increasing the digitiser range, but this would increase the step size Δx and hence the quantisation noise error unless more levels were used. Similar ranging errors occur of course in analogue measurements unless the amplitude bandwidth of the instrument exceeds the input signal's amplitude range. Such errors can be particularly serious in the case of measurement of the higher-order moments such as the skewness or kurtosis. The amplitude levels which contribute most strongly to the integral $\int x^m p(x)\,dx$ rise rapidly as m increases (for Gaussian signals at least) and so particular care in appropriate scaling of the input signal is required if higher moments are desired. This is discussed more fully later.

It is worth pointing out that, in many circumstances, quantisation errors can be significantly reduced by using non-linear step sizes (i.e. by making $\Delta x_i = f(x_i)$). The optimum $f(x_i)$ variation will clearly depend on the particular form of the probability distribution of the signal, so that an arrangement which is beneficial in some circumstances will be detrimental in others. Non-linear encoders are usually only used in rather specialised applications; by far the most common analogue-to-digital systems employ uniform step sizes and the non-uniform case will not be discussed further. It should also be noted that the results given above rest on the assumption that the step size is small compared with the total amplitude range, for only then can $p(x_i)$ be considered constant over Δx_i (unless of course the signal itself has a uform $p(x)$ when no such restriction applies).

In some ways, this 'classical' quantisation error is misleading. The result deduced above, that $\sigma_n^2 = \Delta x^2/12$, does *not* imply that the mean square of the digitised signal exceeds that of the analogue signal by the amount σ_n^2. Consider, for example, the simple case of a signal with a uniform $p(x)$ of $1/2a$ between $-a \leqslant x \leqslant a$ and zero for $|x| > a$ and assume that it is digitised into just two levels corresponding to amplitudes $\pm a/2$. The situation is illustrated in figure 4.1. In this case $\Delta x = a$ and it is easy to see that the measured mean square is given by

$$\sigma_m^2 = \int_{-a}^{a} x^2 p(x)\,\mathrm{d}x = \left(\frac{a}{2}\right)^2 \frac{1}{2a}a + \left(\frac{-a}{2}\right)^2 \frac{1}{2a}a = \frac{a^2}{4}$$

compared with the true mean square of $a^2/3$. So the value measured

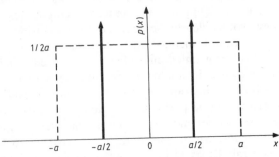

Figure 4.1. 1-bit (two-level) quantisation of a signal having a uniform $p(x)$ of $1/2a$ up to $|x| = a$.

from the digitised signal is actually *lower* than the true value. This simple case can be easily extended to show that, for quantisation to n levels, the measured mean square is given by

$$\sigma_m^2 = \frac{a^2}{3} \frac{n^2 - 1}{n^2}$$

implying that a 4-bit (16-level) ADC would be needed to ensure an error in σ_m^2 of less than 1%.

Similar analysis is possible for signals with different probability density functions although the algebra becomes tedious. In the case of a Gaussian signal with zero mean and unit variance digitised using an n-level quantiser whose dynamic range is $2a$, such that the first quantiser level corresponds to an amplitude defined by $-a + a/n$ (figure 4.2), it can be shown that the measured mean square is given by

$$\sigma_m^2 = a^2 \sum_{i=1}^{n/2} \left(\frac{2i - 1 - n}{n}\right)^2 \left[\left\{1 - \text{erf}\left[\frac{a}{\sqrt{2}}\left(1 - \frac{i}{n}\right)\right]\right\} - 2p_{i-1}\right]$$

where

$$p_{i-1} = 0.5\left\{1 - \text{erf}\left[\frac{a}{\sqrt{2}}\left(1 - \frac{i - 1}{n}\right)\right]\right\}.$$

This result implies an *increase* in σ_m^2 as the number of quantiser levels decreases, in contrast with the previous case of a signal with uniform

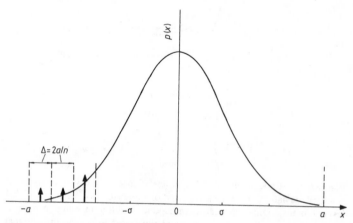

Figure 4.2. n-level ($\ln n/\ln 2$-bit) quantisation of a signal having a Gaussian $p(x)$. The digitiser bandwidth is $2a$.

$p(x)$. It should also be noted that the measured variance depends on both the digitiser bandwidth $2a$ and the number of levels. This is true for any $p(x)$ but was ignored in the previous example by assuming that the digitiser and analogue signal bandwidths were identical.

Figure 4.3 shows how the measured variance of a digitised Gaussian signal varies with the number of quantising levels and the ratio

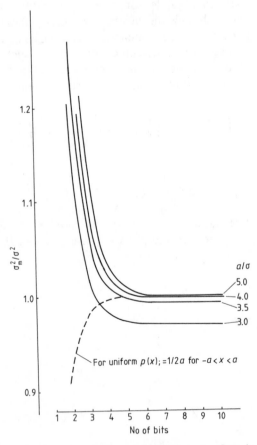

Figure 4.3.   Ratio of measured to true variance of a Gaussian signal of standard deviation σ, using a digitiser of bandwidth $2a$ and up to 1024 levels (10 bits). Different curves correspond to different ratios of a/σ, as shown. Note the different behaviour of the errors in the case of a signal having a uniform $p(x)$.

of the digitiser bandwidth to the standard deviation of the analogue signal. The results given above have been used and the figure includes the result for a uniform $p(x)$ signal. Note that for large enough values of n the results must approach asymptotically those obtained by simple 'analogue' integrations of $\int x^2 p(x)\,dx$ over the specified range. Even if the digitiser bandwidth is six times the standard deviation of the analogue signal (so that the extreme values are where $p(x)$ has fallen to less than 1% of its peak value), the variance is underestimated by about 3%. Even greater errors occur in the higher-order moments as indicated in §4.1. This is demonstrated in figures 4.4 and 4.5 which show the results of similar calculations for the errors in digitised measurements of the fourth and the sixth moment, respectively, for various values of n and a. For Gaussian signals the results imply that errors of less than 1% in digital measurements of the signal variance require at least a 5-bit (32-level) digitiser with a bandwidth in excess of seven times the signal variance; similar accuracy in measurement of the higher moments generally requires more levels.

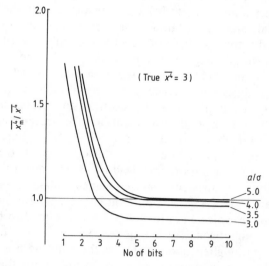

Figure 4.4.   As for figure 4.3, but for the fourth moment of $p(x)$.

Now these results have all been obtained by assuming that signal amplitudes outside the digitiser bandwidth are measured as zeros (i.e.

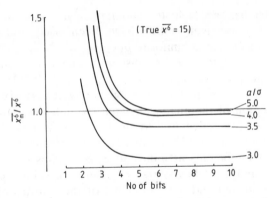

Figure 4.5.   As for figure 4.3, but for the sixth moment of $p(x)$.

the measured $p(x)$ for $|x| > a$ is zero). In practice, this will rarely be true. Even in the case of analogue measurements, if the instrument's input range is exceeded occasionally, the maximum voltage, say, might be held until the signal amplitude falls again—although the details of the resulting distortion on $p(x)$ will depend on the particular electronic characteristics of the device. In the case of a digitised signal, analysis is rather easier since then any signal amplitudes above the top quantisation level will simply contribute to the $p(x)$ value corresponding to that particular level. The effect of the resulting 'spike' in $p(x)$ at $x = a$, the area under which must equal $\int_a^\infty p(x)\,dx$, is always to reduce the errors resulting from simply ignoring $p(x)$ for $x > a$. It is straightforward to include this effect in analysis of the errors in measurements of the various moments of $p(x)$. Figure 4.6 shows results for a Gaussian signal where the percentage error is plotted as a function of the moment (i.e. the m in $x^m = \int x^m p(x)\,dx$) for various bandwidths. These are asymptotic results for analogue signals (or digitised signals obtained using a large number of levels, $n > 256$, say, an 8-bit converter). Results from the earlier analysis in which $p(x)$ for $|x| > a$ was assumed to be zero are included for comparison. As mentioned earlier, errors can become very large for the higher moments unless the bandwidth is sufficiently wide. In practice, it is only in specialised applications that moments above the fourth (kurtosis) are required; so a bandwidth of $(7–8)\sigma$ will usually suffice.

Similar calculations are possible for other forms of $p(x)$ and the analysis can be generalised to cases in which there are additional errors caused by an insufficient number of quantising levels. It should also be emphasised that only signals with a symmetric probability distribution and whose $p(x)$ sits in the centre of the instrument's amplitude bandwidth have been considered; extensions to the analysis are possible to remove these restrictions. However, the calculations rapidly become very tedious and the results discussed above should be sufficient to enable the reader to grasp the major sources of error arising from digitising a signal using a finite number of levels spanning a finite amplitude range. Provided that proper care is taken—and this usually requires *some* prior knowledge of the signal characteristics—these errors can almost always be made small enough to be insignificant. They will be ignored in all the following considerations.

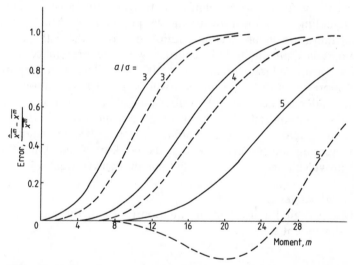

Figure 4.6. Error in measured moments of $p(x)$ due to inadequate amplitude bandwidth $2a$. Quantisation errors are neglected. All results were obtained assuming that $p(x) = 0$ for $|x| > a$ but the broken curves are for cases in which the 'spikes' of area

$$\int_{a}^{\infty} xp(x)\, dx,$$

located at $|x| = a$, are included; full curves assume that $p(x) = 0$ at $|x| > a$.

⟩4.3   Finite sample size errors

In §4.2 the third of the basic questions stated in §4.1—'What digitiser resolution is required?'—was addressed. In this section, we concentrate on the first—'How many individual digitised data samples are required?' Discussion begins with the simplest case of a digital signal containing N *statistically independent* sample values. This immediately implies either that the analogue signal has the spectral characteristics of white noise (uniform energy density at all frequencies and hence zero autocorrelation function $R(\tau)$ for $\tau > 0$) or that it has been digitally sampled either at a rate lower than the lowest-frequency component present or at just those intervals (of τ) for which $R(\tau) = 0$. These are serious restrictions in many cases and discussion of the implications arising from their removal is given later (see §4.4).

Very many textbooks have been written on probability theory and its implications for the assessment of statistical errors arising from the finite sampling of known or unknown populations. Much of this classical literature involves consideration of cases where the number of samples is quite limited, in contrast with the present context in which it is usually large (greater than 100, say). The more important results appropriate to this latter case will be given here without proof and the reader is encouraged to study standard texts if more detailed information is required. (The book by Miller and Freund (1977) is a typical introductory text in probability and statistics for students in the engineering and physical sciences; the book by Kendall and Stuart (1977) provides a more advanced and comprehensive treatise.)

4.3.1 Mean values

An obvious digital equivalent of equation (3.7) for the mean value $\bar{x}$ of N independent samples is

$$\hat{\bar{x}} = \frac{1}{N}\sum_{i=1}^{N}x_i. \tag{4.2}$$

Now a second set of N digital samples from the same analogue signal (or, in the common probability parlance, 'population') would not generally yield the same $\hat{\bar{x}}$. In that sense the right-hand side of equation (4.2) is simply an 'estimator' and $\hat{\bar{x}}$ is itself a random variable. One of the basic sampling theorems states that a random sample of size N taken from a (large) population of mean $\bar{x}$ and variance σ^2 yields a random variable $\hat{\bar{x}}$ whose mean value is $\bar{x}$ and whose variance is σ^2/N. The fact that the variance of the $\bar{x}$ estimates is σ^2/N is significant, for it clearly implies that the difference between $\hat{\bar{x}}$ and the true mean can be made as small as necessary by choosing a suitably large number of samples. Note that to reduce the likely error by a factor of, say, 2 requires a fourfold increase in N.

Complete statistical information about $\hat{\bar{x}}$ is only possible if the distribution of x_i itself is known or if, as a limiting case, $N \to \infty$. In the present context this is not a problem since, although the distribution of x_i is *not* usually known, it *is* almost invariably the case that $N \gg 1$. Then the central-limit theorem can be stated in the form that, if $\hat{\bar{x}}$ is the mean of a random sample of size N taken from a population having mean $\bar{x}$ and (finite) variance σ^2, then

$$z = (\hat{\bar{x}} - \bar{x})/(\sigma/\sqrt{N})$$

is a random variable whose distribution function approaches the normal distribution as $N \to \infty$.

It should be emphasised that this is true regardless of the form of the distribution of x_i, which is what makes this theorem so important in statistical theory. In the particular case of a normally distributed population (of x_i), it can be shown that z is exactly normal for *any* N. It is also worth noting that, even if the x_i population distribution is far from normal, surprisingly small values of N still yield a z population quite closely normal. This is illustrated in figure 4.7 where, as an example, 40 values of z, each obtained by finding $\hat{\bar{x}}$ using just 25 samples of a *uniformly* distributed x_i population, are plotted cumulatively and compared with the normal cumulative distribution $(\int_{-\infty}^{z} \exp(-t^2/2) \, dt)$. Figure 4.8 shows the same data plotted in the more usual form of a confidence limit chart (see below). Note that these data were obtained using the software package, as described in §4.6.2; the details given in that section should be sufficient to allow the user to generate similar data of his or her own.

It will be apparent that the results given above can strictly only be obtained if the variance of the x_i population is known. This does not pose any difficulties if N is large, for then it is reasonable to use the estimated value of σ denoted $\hat{\sigma}$ (see below). For small values of N, little is known about the statistics of the variable $(\hat{\bar{x}} - \bar{x})/(\hat{\sigma}/\sqrt{N})$ unless the x_i are normally distributed, but, if this is the case, then this variable has the 'Student t' distribution, as discussed in standard texts. It can be shown that this distribution approaches the normal distribution as $N \to \infty$. In practice, the differences are insignificant for $N > 30$, so for the present purpose discussion is limited to the sampling variable z.

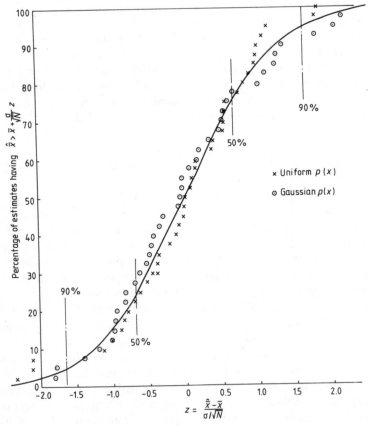

Figure 4.7. Cumulative distribution of 40 estimates of the mean value, each obtained using 25 samples: ×, uniform $p(x)$, 21 estimates inside the 50% confidence interval, five estimates outside the 90% confidence interval; O, Gaussian $p(x)$, 20 estimates inside the 50% confidence interval; five estimates outside the 90% confidence interval. 'True' values of $\bar{x}$ and σ are obtained using all 1000 samples.

There are two desirable properties of any estimator of a particular statistic such as equation (4.2) for the mean value. Firstly, it is usual to require that the mean of its sampling distribution is equal to the

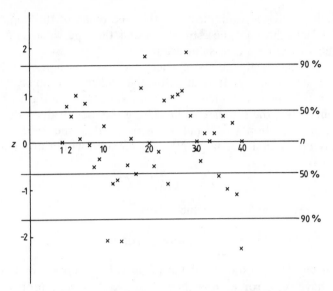

Figure 4.8. Confidence limit chart for random (uniformly distributed) data for 40 estimates of the mean, each using 25 samples.

true value of the original population's statistic. If this is so, the estimates are said to be unbiased. For example, in the context of mean value estimation, we want the mean of $\hat{\bar{x}}$ to be equal to $\bar{x}$. It can be shown that this *is* true if equation (4.2) is used to obtain the $\bar{x}$ estimates, so this obvious estimator is in fact unbiased. However, other unbiased estimators are available as an estimate of $\bar{x}$. One example is the midrange value—the mean of the largest and smallest values in the sample. It is therefore clearly necessary to use a further criterion for deciding which of the estimators is best in a particular case. This is provided by the variance of the sampling distribution. The best estimator will usually be the one which leads to the smallest variance in the estimates. We have already seen that the variance of $\hat{\bar{x}}$, if the latter is obtained using equation (4.2), is σ^2/N (see, e.g., Kendall and Stuart 1977). Now provided that the original population's distribution is symmetric, it can be shown that the variance of $\bar{x}$ estimates obtained by using the middlemost (median) values of each sample set (i.e. the value of the $(N/2)$th sample when they are arranged in numerical order) is about $1.57\sigma^2/N$. Equation (4.2) is therefore a more efficient estimator of $\bar{x}$ than the median value, since it is likely to give a result closer to the true mean of the population. It is also desirable that the estimator yields the true value as $N \to \infty$; it is then said to be a consistent estimator.

For any single estimate the chances are virtually non-existent that $\hat{\bar{x}}$ will exactly equal $\bar{x}$. It is therefore helpful to accompany such an estimate with some statement concerning how close to the true value we might expect the estimate to be. Using the mean-value estimate as defined by equation (4.2) and recognising that the resulting z is normally distributed (for large N), we can state with a probability of $1 - \alpha$ that

$$-z_{\alpha/2} < (\hat{\bar{x}} - \bar{x})/(\sigma/\sqrt{N}) < z_{\alpha/2} \qquad (4.3a)$$

where $z_{\alpha/2}$ is the value of z which gives

$$\frac{1}{\sqrt{2\pi}} \int_z^\infty \exp\left(\frac{-t^2}{2}\right) dt = \frac{\alpha}{2}.$$

In other words, an estimate of the mean value from a sample of size N will have an error ε_x less than $z_{\alpha/2}\sigma/\sqrt{N}$ with a probability of $1 - \alpha$. To put it another way, an error ε_x with a certainty of $1 - \alpha$ requires a sample size given by

$$N > (\sigma z_{\sigma/2}/\varepsilon_x)^2. \qquad (4.4a)$$

It is sometimes convenient to use a normalised error, defined by $\varepsilon_x' = \Delta x/\bar{x}$ ($\Delta x = \hat{\bar{x}} - \bar{x}$), in which case

$$N > [(\sigma/\bar{x})(z_{\alpha/2}/\varepsilon_x')]^2. \qquad (4.4b)$$

Obviously this is only useful if $\bar{x} \neq 0$. The implications of these statements for N should be intuitively obvious. If we wish to increase the certainty with which the statement can be made (i.e. to increase $z_{\alpha/2}$) or to reduce the likely error in the estimate, then more samples are required. Similarly, more samples are required to achieve the same likely error for a population with a high variance than for one with a low variance.

It is common to restate equation (4.3a) in terms of a confidence interval, writing

$$\hat{\bar{x}} - z_{\alpha/2}\sigma/\sqrt{N} < \bar{x} < \hat{\bar{x}} + z_{\alpha/2}\sigma/\sqrt{N}. \qquad (4.3b)$$

This represents the claim that with a probability (or confidence) of $1 - \alpha$ the interval from $\hat{\bar{x}} - z_{\alpha/2}\sigma/\sqrt{N}$ to $\hat{\bar{x}} + z_{\alpha/2}\sigma/\sqrt{N}$ contains the true mean value $\bar{x}$. As an example the 50% and 90% confidence intervals, corresponding to $z_{0.25}$ and $z_{0.05}$ (0.674 and

1.645, respectively) are included in figures 4.7 and 4.8. Note that, whilst on average we would expect four out of the 40 estimates to lie outside the 90% confidence limit, it happens that five estimates do so in both the Gaussian and the uniform $p(x)$ case. In the latter case, one might generally anticipate fewer than four estimates, recognising that the maximum value of a *single* sample is only $\sigma \sqrt{3}$ (since $\sigma^2 = a^2/3$ for uniform $p(x)$ between $\pm a$); this provides a definite upper bound on $\hat{\bar{x}}$, unlike the case of a Gaussian $p(x)$ for which x_i can take any value.

A common and related way of judging the reliability of an estimate of the mean, or any other statistical parameter, is to use the 'standard error'. This is normally defined simply as the square root of the variance of the sampling distribution of the estimates. In the case of mean-value estimates, since the variance of $\hat{\bar{x}}$ is σ^2/N the standard error is just $\sigma/\sqrt{N}$ or, provided that N is large, $\hat{\sigma}/\sqrt{N}$. Specific confidence intervals are then seen as *quantitative* statements concerning the likelihood that the true mean $\bar{x}$ lies within a specific number of standard errors $z_{\alpha/2}\sigma/\sqrt{N}$ either side of the estimate $\hat{\bar{x}}$.

In the context of signal analysis, it is usually sufficient to note the result expressed by equation (4.4) and to ensure that N is large enough to provide a suitably small error with a sufficiently high certainty. It should be emphasised that, provided that $N > 30$, say, equation (4.4) does *not* depend on the fact that the original population is normally distributed. The single and major requirement is that the samples used should be statistically independent, as mentioned earlier. In §4.4, we discuss what happens if they are not.

4.3.2 Mean-square values In the case of the signal variance, an obvious digital equivalent of equation (3.8) (expressed as the variance) is

$$\hat{\sigma}^2 = \frac{1}{N}\sum_{i-1}^{N}(x_i - \hat{\bar{x}})^2. \qquad (4.5)$$

This is not, in fact, an unbiased estimator; one fairly obvious reason is that there are only $N - 1$ independent deviations of x_i from the mean, since their sum is always zero. $N - 1$ values of $x_i - \hat{\bar{x}}$ automatically determine the Nth value. In most statistical texts it is therefore common to use $N - 1$ as the divisor instead of N, but this is an unnecessary nicety in the present context, where N is invariably large.

Now, since $\hat{\sigma}^2$ is inevitably positive, it is immediately evident that its distribution cannot be normal. The actual distribution is called the chi-squared distribution where, specifically

$$\chi^2_{N-1} = (N - 1)\hat{\sigma}^2/\sigma^2.$$

Like the Student-t distribution, that of χ^2 depends in general on $N - 1$ (the number of degrees of freedom) but, if N is large enough (greater than 30), it is closely normal. It should be emphasised, however that the sampling distribution of the variance estimates is only χ^2 if the x_i population is normal. Confidence statements regarding the likely error in σ estimates can in this case be made in a way similar to those for $\bar{x}$ estimates. Thus the true value of σ^2 obtained will, with a probability of $1 - \alpha$, lie in the range

$$(N - 1)\sigma^2/\chi^2_{N-1,\alpha/2} < \hat{\sigma}^2 < (N - 1)\,\sigma^2/\chi^2_{N-1,1-\alpha/2}$$

if equation (4.5) (with $N - 1$ as the divisor) is used to obtain $\hat{\sigma}$. Note that the two χ^2 values differ since the distribution is not symmetric. In the case of large N, this can be simplified, for then the series expression for χ^2 (see standard texts) can be reduced to

$$\chi^2_{N,\alpha} = N + z_\alpha\sqrt{2N}.$$

Hence the $1 - \alpha$ confidence interval (now symmetric) is given by

$$\sigma^2(1 - z_{\alpha/2}\sqrt{2/N}) < \hat{\sigma}^2 < \sigma^2 (1 + z_{\alpha/2}\sqrt{2/N}). \qquad (4.6)$$

The result for the required number of samples corresponding to that for the mean-value estimate (equation (4.4)) is then just

$$N > 2(z_{\alpha/2}\sigma^2/\varepsilon_\sigma^2)^2 \qquad (4.7a)$$

where $\varepsilon_\sigma = \hat{\sigma}^2 - \sigma^2$ or, in terms of a normalised error defined by $\varepsilon'_\sigma = (\hat{\sigma}^2 - \sigma^2)/\sigma^2$,

$$N > 2z^2_{\alpha/2}/\varepsilon_\sigma'^2. \qquad (4.7b)$$

Note that, as would be expected, this is generally more restrictive for a given accuracy than the corresponding requirement for mean value estimates. Figure 4.9 shows the cumulative distribution of 40 estimates of σ^2, each obtained using a different set of 100 samples from a uniform and a Gaussian population. (Again, the software can be used to generate these or similar results (see §4.6.2).) It should be noted in particular that the results for the uniform population are clearly quite different from the normal cumulative distribution, emphasising that, unlike the case of mean-value estimation, the distribution of $\hat{\sigma}^2$ is not normal for large N (i.e. equation (4.6) is not valid) unless the original population is itself normal. The deviation from normality will of course depend on the particular form of the probability distribution of the original population. The variance of the mean-square estimates is given by

$$\text{var}(\hat{\sigma}^2) = \sigma^4/(N/2). \qquad (4.8)$$

The standard error for variance estimates is therefore $\sigma^2\sqrt{2/N}$.

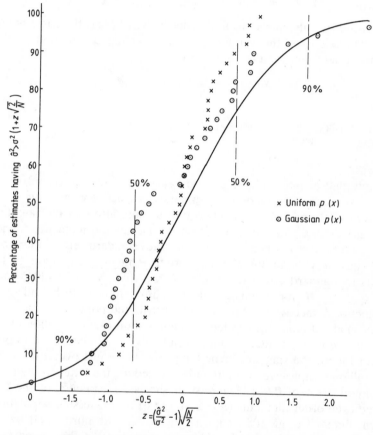

Figure 4.9. Cumulative distribution of 40 estimates of the variance, each obtained using 100 samples: ×, uniform $p(x)$, 30 estimates inside the 50% confidence interval, no estimates outside the 90% confidence interval; ○, Gaussian $p(x)$, 16 estimates inside the 50% confidence interval, three estimates outside the 90% confidence interval. σ was obtained using all 4000 samples.

4.3.3 Higher-order moments

The obvious digital equivalent of equation (3.10) for calculation of higher-order mean moments of a digitised signal is

$$\widehat{\overline{x^n}} = \frac{1}{N}\sum_{i=1}^{N}(x_i - \widehat{\overline{x}})^n. \tag{4.9}$$

It can be shown that the variance of the sampling distribution is given by

$$\text{var}(\widehat{\overline{x^n}}) = (1/N)[\overline{x^{2n}} - (\overline{x^n})^2 + n^2\sigma^2(\overline{x^{n-1}})^2 - 2n\overline{x^{n-1}}\,\overline{x^{n+1}}]. \tag{4.10}$$

This result is only correct to first order in N, whereas the corresponding expression for the variance of the sampling distribution of the higher-order moments themselves, i.e.

$$\text{var}(\widehat{\overline{x^n}}) = [\overline{x^{2n}} - (\overline{x^n})^2]/N$$

with

$$\widehat{\overline{x^n}} = \frac{1}{N}\sum_{i=1}^{N}x_i^n$$

is exact.

Note that equation (4.10) gives variances of σ^2/N and $(\sigma^2)^2 2/N$ for the first- and second-moment estimates, respectively, as anticipated from §§4.3.1 and 4.3.2. It is important to note also that the sampling variance of any moment depends on the population moment of twice the order. Even when N is large the standard error therefore becomes very large for the high-order moments. Equation (4.10) leads to standard errors of $\sigma^3\sqrt{6/N}$, $\sigma^4\sqrt{96/N}$, $\sigma^5\sqrt{720/N}$ and $\sigma^6\sqrt{10\,170/N}$ for estimates of the third, fourth, fifth and sixth moments, respectively, of a normally distributed population. In terms of normalised moments (γ, defined by equation (3.9)), it follows that, whilst 10 000 independent samples would give a standard error of only 0.014 for γ_2, the standard error for γ_6 would be just over 1.00. Over 50 million samples would be required to reduce the latter to the same standard error of 0.014! Figure 4.10 shows 20 estimates of the first three even-order moments ($n = 2$, 4 and 6) of a Gaussian population, each obtained using 100 samples. The results are normalised by the exact values and the rapidly increasing variability of the estimates as n increases is evident. In this case the data were sampled from the Gaussian (white) noise population available in the computer package. Each sample could take values between -127 and 127 (1-byte data)

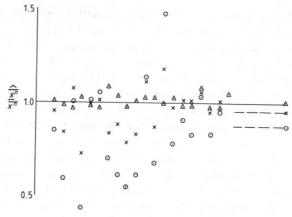

Figure 4.10. Even-order normalised moments from sets of 100 samples of a Gaussian population (8-bit *data*, with σ such that $3\sigma = \pm127$): $\triangle$, $n = 2$; $\times$, $n = 4$; $\bigcirc$, $n = 6$; chain lines show corresponding results for 19 456 samples. Note that for the fourth and the sixth moment the latter are not equal to unity because of ranging errors introduced by making $3\sigma = 128$.

and the standard deviation of the whole set (19 456 samples) was such that the extreme values (±127) corresponded to about $\pm3\sigma$. As would be anticipated from the discussion in §4.2, this leads to inevitable ranging errors in the high-order moment calculations, as can be seen from the results in figure 4.10. The fourth and sixth moments of the complete data set were only 2.883 and 13.15 instead of the exact Gaussian values of 3 and 15, respectively. For the normalisation used in the figure the standard errors for these two moments are expected to be about $(1/3)\sqrt{96/100} = 0.33$ and $(1/15)\sqrt{10\,170/100} = 0.67$, respectively; the largest variations in the results from the 19 456 samples do happen to have roughly these values. These results all emphasise the difficulty of obtaining high-order moments with good accuracy, even in the absence of the ranging errors discussed in §4.2.

As in the case of mean and variance estimates the standard error result expressed by equation (4.10) is quite independent of the probability distribution of the original population but obviously, whether or not the latter is Gaussian, the sampling distribution of $\widehat{x^n}$ will only tend to normality for exceedingly large N—increasingly large as n increases. This should be obvious intuitively; the major contributions to the high-order moments arise from sample values remote from the mean and these occur relatively infrequently. The sampling distribution can be expressed as $z = (\overline{x^n}/\overline{x^n} - 1)(N/F)$, where F is the factor arising from the right-hand side of equation (4.10)—10 170 for $n = 6$ and a Gaussian parent distribution. Since normality of z implies that z can take, in the limit, *any* value, it is clear that very large N values are needed to ensure that the sample values remote from the mean are properly represented.

This is one of the basic reasons why consideration of higher-order moment measurements is often not included in statistical textbooks. In most applications of sampling theory the number of samples possible to acquire in practice is quite limited; so such measurements are impossible to obtain with reasonable accuracy. In the context of signal analysis, however, there are occasions when higher-order moments are required in order, say, to help in the interpretation of the underlying physical phenomena giving rise to the signal. Evidently in such cases very large sample sizes must be used—equivalent of course to very long analogue integration times. The computer exercises included in §4.6.3 will give the reader further insight into the difficulties which arise in making measurements of high-order moments.

4.3.4 Probability density estimates

In some applications, it is necessary to measure the complete probability density function of the signal rather than just some of the moments. This requires estimation of $p(x)$ across the whole amplitude range of the signal and the obvious digital equivalent of equation (3.1) is

$$\hat{p}(x) = N_x/(\Delta x \, N)$$

where N_x is the number of samples (out of the total number N) which have an amplitude in the range $x - \Delta x/2$ to $x + \Delta x/2$. Estimation of the complete $p(x)$ distribution therefore requires prior decisions concerning not only the number of samples to be used but also the amplitude range to be covered and the number of $p(x)$ estimates, M say, required in that range. If the signal is estimated to spread over a range $\pm R$, say, then Δx will be $2R/(M - 1)$, assuming that Δx (the 'slot' width) is constant across the range.

Theoretical deductions about the sampling distribution of $\hat{p}(x)$ are very difficult to obtain, even if $p(x)$ is normal, but some features of the variability of the estimates should be immediately obvious. As usual, one expects the standard error to decrease as N increases. Generally, it would also be anticipated that any set of parameters that leads to a higher N_x/N ratio in a given slot will lead to smaller variability in that $\hat{p}(x)$ estimate. An obvious way to improve the likely accuracy is therefore to increase Δx but this naturally leads to less resolution across the amplitude range (less slots). Further, it should be clear that, if Δx is kept constant for all x, the variability will be higher in regions of x where $p(x)$ is relatively small; in these regions, relatively few samples occur (N_x/N is small) and so there is greater margin for error.

These intuitive deductions are embodied in the approximate result for the variability of $p(x)$ estimates:

$$\text{var}[\hat{p}(x)] \simeq p^2(x)/[N \,\Delta x\, \hat{p}(x)]. \tag{4.11}$$

As an example of the way in which the above expression can be used to obtain an estimate of the number of samples required to measure a complete $p(x)$, consider the case of a normally distributed population with zero mean and unit variance. Assume that 32 $p(x)$ estimates are required across an amplitude range given by $\pm 3\sigma$ and that the standard error is to be kept below 5% of the peak $p(x)$. Using a standard error normalised by the peak $p(x)$, equation (4.11) leads to the approximate expression

$$\varepsilon' = \varepsilon_{p(x)}/p(0) \simeq \sqrt{p(x)/(N\,\Delta x)}\,/p(0).$$

So, for a Gaussian distribution with M slots covering an amplitude range of 6σ,

$$N = M\sqrt{2\pi}\,/6\varepsilon'^2.$$

For the example above, $M = 32$ and $\varepsilon' = 0.05$, giving $N = 5350$. Reducing ε' to 2% or increasing M to 200 will increase the required number of samples to about 33 000. This example illustrates the point that large sample sizes are required to obtain accurate probability density estimates, particularly if fine amplitude resolution is also required.

The computer examples can be used to study the influence of the various parameters in more detail and figure 4.11 shows some typical results. Probability distribution estimates have been obtained from data sets which nominally have either a uniform or a Gaussian distribution. In both cases, 16 estimates (slots) were obtained with 1000 or 10 000 samples. The decrease in variability for the latter case is clear and the increase in variability arising from increasing the

70 AMPLITUDE-DOMAIN STATISTICS

number of slots (to 64) whilst using the same number of samples (10 000) is also shown in the case of the Gaussian data set. The standard error limits corresponding to this latter case ($\varepsilon' p(0) = 0.021$ at $x = 0$) are included and it is evident that only a few of the $p(x)$ estimates lie outside the nominal error band.

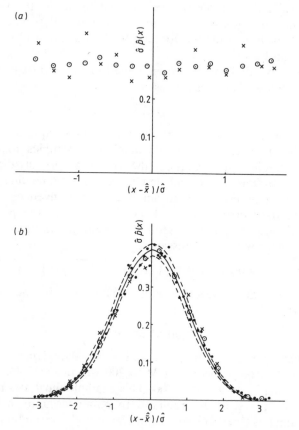

Figure 4.11. Probability density estimates for (a) a uniform population and (b) a Gaussian population: ×, 16 estimates from 1000 samples; ○, 16 estimates from 10 000 samples; ●, 64 estimates from 10 000 samples; ———, $p(x) \pm$ standard error calculated for $M = 64$, $N = 10\,000$, with $\Delta x = 6\sigma/(M - 1)$. The full curve is the exact Gaussian distribution.

It should be emphasised that equation (4.11) provides only a qualitative guide and in practice it is quite common, in the context of signal analysis, to work up confidence in the $p(x)$ estimates by the rather *ad hoc* procedure of arbitrarily varying Δx and N and simply 'seeing what happens'. This process can be illustrated by using the computer options described in §4.6.4.

⟩**4.4 The effects of using correlated samples**

Attention has been concentrated thus far on the statistical errors which arise when using a finite number of statistically independent sample values. However, where the signal being digitised has spectral characteristics different from white noise, consecutive digitised samples may be highly correlated. As discussed in Chapter 5, this is usually the inevitable consequence of using a sampling rate high enough to enable time-domain statistics (autocorrelation or spectra) to be adequately determined. The use of the consecutive samples to obtain amplitude-domain statistics will then generally be rather wasteful, since many more would be required to reduce the statistical errors to those estimated on the basis of uncorrelated samples. In such cases, what is usually required for an efficient estimation of, say, the mean or mean-square value is, first, an effective total sample *time* long enough to average the lowest-frequency content of the signal adequately and, secondly, a sample set sufficiently large to minimise the statistical error arising from a finite sample size. A typical example will serve to illustrate this point.

Consider a sample set generated by digitising a pink noise signal with an autocorrelation given by $R(\tau) = \exp(-\tau/T)$. If the sampling rate were lower than, say, $1/4T$, then consecutive samples would on average be almost independent $(R(4T) < 0.02)$. The ideas discussed in the previous sections would then be appropriate for determining the number of samples required for estimation of a particular amplitude-domain characteristic. These samples would not allow the informative part of the autocorrelation to be recovered $(\tau < 4T)$. If adequate measurements of the latter were required, it would clearly be necessary to choose a much more rapid sampling rate of $- 1/0.1T$, say. (In Chapter 5, we discuss the limitations on measurement of $R(\tau)$ and/or $E(f)$ arising from particular sampling rates.) Consecutive samples would then be highly correlated $(R(0.1T) = 0.9)$ and it is obvious that the number of samples required to obtain adequate amplitude-domain statistics would be much greater than in the case of the lower sampling rate.

Figure 4.12 shows estimates of the mean value of a pink-noise signal having a zero mean and a unit variance. Each estimate was calculated using 100 samples and three sets of estimates are included, obtained using effective sampling rates of $10/T$, $2/T$ and $1/T$. It is clear that only for the final set do the estimates roughly conform to what would be expected on the basis of the Gaussian statistics of independent samples. For the highest sampling rate $10/T$, an additional set of mean-value estimates are shown, each obtained using 1000 samples so that the total sampling time was the same as it was for the estimates based on 100 samples and a $1/T$ sampling rate. The variability is evidently about the same.

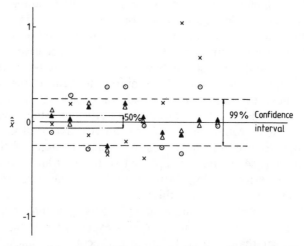

Figure 4.12. Mean-value estimates from 100 ($\times$, $\bigcirc$, $\triangle$) or 1000 ($\blacktriangle$) samples of pink noise with $R(\tau) = \exp(-\tau/T)$: $\times$, $\blacktriangle$, $\Delta\tau = 0.1$; $\bigcirc$, $\Delta\tau = 0.5$; $\triangle$, $\Delta\tau = 1.0$. Confidence intervals refer to $N = 100$.

Once the sampling rate becomes sufficiently high, the variability of the estimates of amplitude-domain statistics is determined essentially by the total sampling time and the spectral nature of the signal. Standard texts (see, e.g. Bendat and Piersol 1966) derive expressions for the variability in these circumstances and the results are equally applicable to analogue measurements. In general, it can be shown that the variability of mean-value estimates is given by

$$\mathrm{var}(\hat{\bar{x}}) = \frac{1}{T}\int_{-T}^{T}\left(1 - \frac{|\tau|}{T}\right)R(\tau)\,\mathrm{d}\tau.$$

Now, provided that $|\tau| \ll T$, this becomes

$$\text{var}(\hat{\bar{x}}) = \frac{1}{T}\int_{-\infty}^{\infty} R(\tau)\,d\tau \qquad (4.12)$$

and this all assumes that the $\hat{\bar{x}}$ values are obtained from

$$\hat{\bar{x}} = \frac{1}{T}\int_{0}^{T} x(t)\,dt.$$

The equivalent expression for the variability of estimates of the signal variance is

$$\text{var}(\hat{\sigma}^2) = \frac{2}{T}\int_{-\infty}^{\infty} [R^2(\tau) + 2\,\bar{x}\,R(\tau)]\,d\tau. \qquad (4.13)$$

Consider, for example, the case of bandwidth-limited white noise having equal power at all frequencies between 0 and B, say, and a variance of σ^2. It follows from the above and the result for $R(\tau)$ given in §3.2.1 that

$$\text{var}[\hat{\bar{x}}] = \frac{\sigma^2}{T}\int_{-\infty}^{\infty} \frac{\sin(2\pi B\tau)}{2\pi B\tau}\,d\tau = \frac{\sigma^2}{2BT} \qquad (4.14)$$

provided that T is sufficiently large. Similarly, the variability of the variance estimates is

$$\text{var}(\hat{\sigma}^2) = \sigma^4/BT. \qquad (4.15)$$

The implication is that, for a given variability, the sampling time must increase as the signal (frequency) bandwidth decreases. This is to be expected, since a small signal bandwidth means that relatively more of the signal energy is at the lower frequencies. Note that, if the signal were sampled at intervals of $1/2B$, yielding truly uncorrelated samples (see §3.2.1), then $2BT = N$, the total number of samples, so that the variability of the digital estimates of the mean value is just σ^2/N, as expected. The reader is reminded here that in the case of white noise ($B = \infty$) the variability depends only on N (and not on T).

Evaluating the variability of probability density estimates is much more difficult, even in cases of particularly simple signals, because it requires a knowledge of the statistical properties of the time intervals between each signal element lying within each amplitude band. This is generally very difficult to ascertain. However, heuristic arguments can be used to deduce that, in the case of bandwidth-limited white noise, the expression analogous to equations (4.14) and (4.15) is

$$\text{var}[\hat{p}(x)] \simeq Ap^2(x)/[BT\,\Delta x\,\hat{p}(x)] \qquad (4.16)$$

where A is a constant of order unity.

In practical terms, since most signals of any physical interest have spectral characteristics quite unlike bandwidth-limited white noise and, furthermore, one often does not know these characteristics *a priori*, the results given in this section should be used with caution. They will generally represent lower bounds on variability estimates. If the autocorrelation funtion *is* known (even approximately), then equations (4.12) and (4.13) may be useful in the case of mean and mean-square measurements. A sensible procedure in most cases, however, is simply to ensure that the effective averaging time is significantly longer than the period of lowest-frequency fluctuations in the signal. This will inevitably require some element of trial and error in the experimental procedure.

⟩4.5   Summary and examples

In this section the more important sampling criteria discussed in earlier sections will be summarised and a few examples of their use given. It is assumed throughout that quantisation errors are negligible.

The standard errors in the measurement of the first six moments of the probability density function, i.e. the mean, the variance and the third to sixth moments, are given by

$$\varepsilon_s = \sigma/\sqrt{N},\ \sigma\sqrt{2/N},\ \sigma\sqrt{6/N},\ \sigma\sqrt{96/N},\ \sigma\sqrt{720/N}$$
$$\text{and } \sigma\sqrt{10\,170/N} \qquad (4.17)$$

respectively. It should be emphasised that these results are only valid when the N samples are statistically independent and, except in the case of the mean value, for populations which are normal. Again, except in the first case, they also refer to the moments of the *fluctuating* part of the data, i.e. with the mean value removed.

In the case of the mean and variance estimates, which are by far the most common, sampling theory shows further that given the above conditions, the mean and variance estimates will lie within an amount ε of their true values with 95% probability for sample sizes satisfying

$$N > 4\sigma^2/\varepsilon_x^2 \qquad (4.18)$$

and

$$N > 8\sigma^4/\varepsilon_{\sigma^2}^2 \qquad (4.19)$$

respectively. Note that in these expressions the ε are not standard errors, but simply the difference between the true and measured values. Note also that $z_{\alpha/2}$ has been taken as 2 rather than the more exact value of 1.96 for 95% confidence. The second of these expressions is only valid for large N (>50, say). Corresponding expressions for the higher-order moments are not generally available because the sampling distributions are very far from normal. However, in practice, a reasonable guide to the required number of samples is provided by simply applying the appropriate factor from the standard error results in equations (4.17).

As an example, suppose that measurements of the first six moments were required with a likely error (normalised by the true value) of less than 5%. For the variance measurement, $\varepsilon_{\sigma^2}/\sigma^2$ of less than 0.05 with 95% probability requires some 3200 statistically independent samples (from equation (4.19). The same level of error on the sixth-moment measurement, however, is likely to require about

$$3200 \times 10\,170/2 \times \sigma^6/\overline{x^6}.$$

For a Gaussian distribution $(\overline{x^6}/\sigma^6 = 15)$ this is about $1\,084\,800$ samples. Taking only 3200 samples for the sixth moment may lead to normalised errors as high as $0.05 \times \sqrt{10\,170/2}/15 = 24\%$ with 95% probability. The standard error would be about 12%.

In the case of measurement of the probability density function an approximate expression for the standard error, corresponding to equation (4.16) but normalised by the peak $p(x)$ (i.e. $p(0)$), is

$$\varepsilon_s = [p(x)/(N\,\Delta x)]^{1/2}/p(0).$$

Assuming that M $p(x)$ estimates are required across an amplitude range of $\lambda\sigma$, where λ may vary for different applications, this becomes approximately

$$\varepsilon_s = \sqrt{(M/N)(2\pi\sigma/\lambda)p(x)}.$$

The maximum value of this error occurs at $x = 0$ and is thus

$$\varepsilon_{s\,max} = \sqrt{(M/N)(\sqrt{2\pi}/\lambda)}$$

leading to a convenient approximate expression for the required number of samples:

$$N = 2.5M/\lambda\varepsilon_{s\,max}^2. \qquad (4.20)$$

The example given in §4.3.4 was for the case when $\lambda = 6$, $M = 32$ and $\varepsilon_{s\,max} = 0.05$, which leads to a required sample size of about 5350.

It must be emphasised again that equations (4.17)–(4.20) can only provide rough guidelines; none of these equations is correct for non-Gaussian data and only equation (4.17) is exact for Gaussian data (apart from the stated rounding up of $z_{\alpha/2}$). Further, if the samples are not statistically independent, these results will *underestimate* the number of samples required for a specified accuracy. They should therefore always be used with caution.

⟩4.6 Computer exercises

Demonstrations and exercises appropriate to the preceding sections can be undertaken by running CHAP4 from the main index. The screen first clears and then displays a further index, referred to later in the action prompts as the "CHAP4 index". After selecting any one of the first three options in the usual way, the screen clears and then an appropriate title appears, together with the two highlighted option windows, containing information on the currently selected data set and the printer status as described in Appendix B. It is suggested that the user selects each of the three program options in turn and studies the various facilities available within each routine by using, initially, the Gaussian (white) noise data set. Once the latter has been selected (by using the SPACE bar) the RETURN key will continue the program, as usual.

The fourth option of 'display any signal' allows a dynamic display of any of the available data sets. Selecting this option leads first to a request for the disc drive number on which the required data set is resident (two by default) and then a list of all the available data sets on that disc is displayed. Unless the user has created further sets of his own (see §4.6.1 and Appendix A), the list will contain simply the four supplied files. The user may select any one of these by entering the appropriate number. After the data set has been loaded into the memory, its individual values are plotted (vertically) against a convenient horizontal scale which can therefore be thought of as equivalent to an oscilloscope time base. Note that, in the case of the BBC software, mode 4 is used for this graphic display and so only the first

10 000 or so data values are available (mode 4 memory runs from &5800). This option is very useful for reminding the reader of the visual appearance of the signal; screen dump files can be obtained by the appropriate response to the action prompts.

In §§4.6.2–4.6.4 the main three options are described, but we start with a brief description of the four available simulated signals.

4.6.1 The simulated signals

Whilst remarks made in this section are largely in the context of the BBC software, a similar approach was taken for the PC version of the package. Each of the simulated signals consists of a set of 19 456 1-byte data values, stored on the system disc but loaded into memory when required. The random-data set (DAT0N8) was initially produced using the BASIC random-number generator. Numbers were normalised to have values of between −127 and 127 (8 bits) and some minor adjustments made to ensure a mean value over all 19 456 samples of near zero. In common with most versions of BASIC, the BBC's BASIC random-number generator produces numbers between specified limits (zero and unity, typically) with uniform probability of occurrence; so this first data set nominally has a uniform $p(x)$ for $-127 < x < 127$.

The Gaussian data (DAT1N8) were formed by suitable arithmetical operations on random data. Standard techniques were used and are outlined in Appendix A. It should be noted, however, that the data are normalised so that the standard deviation is about 40 (with a mean near zero); this means that the $\pm 3\sigma$ points lie just inside the amplitude bandwidth corresponding to 8-bit data (± 127). The extreme 'tails' of the probability distribution are therefore not captured in the data and, as discussed in earlier sections, this leads to inevitable ranging errors, particularly in the calculation of the higher-order moments. The user may wish to provide alternative data sets of his or her own; this should be quite straightforward (using the information given in the Appendix A) and if such a data set were saved on a copy of the system disc using the filename ":2.$.DAT0N8", it would be loaded when the user requests the random data, since that has the same filename.

For the signals simulating phenomena with a genuine structure in time, standard first- and second-order filtering techniques were applied, using a Gaussian (white) noise data set. Ideally these (linear) processes should not alter the probability distribution of the original signal, so that both the pink noise signal and the second-order signal should have Gaussian statistics. However, as the user will no doubt discover for himself or herself, the probability density distributions of these signals are both slightly skewed. This is partly a result of the limited number of samples used, but more importantly a result of numerical errors in the filtering operations (see Appendix A). For the pink-noise data set (DAT2N8) a signal having $R(\tau) = \exp(-\tau)$ with sample increments of $\Delta\tau = 0.1$ was simulated and for the second-order signal (DAT3N8) a signal having $R(\tau) = \cos(\pi\tau)\exp(-\tau)$ was used with a 0.025 sample increment. In both cases the mean values and the standard deviations are again nominally zero and 40, respectively.

In many of the options described in the following sections the user is prompted for the number of samples (NS) required for a calculation, the number of blocks (NB) of such samples, the sampling increment (INC) and the position of the first sample in the first block (D). The sampling increment determines whether or not consecutive samples from the original set are to be used; the default value is normally one. If higher values are used (n, say) then only every nth sample is used. This parameter is therefore equivalent to a sampling *rate*. If more than one estimate of the required parameter (like the variance) is required then NB > 1 and the blocks run consecutively. Since only 19 456 samples are available there are obvious limits on these various parameters, determined by D + NS*INC*NB<19 456. Attempts to insert values which would require more than 19 456 values are trapped.

4.6.2 Mean and mean-square values
There are two basic routines in this option, which are selected from action prompts after the initial pressing of RETURN. The first ('Confidence limit demonstrations') allows mean value estimates to be displayed graphically in the form of confidence limit charts. If this option is selected, an initial demonstration is presented in which a contiguous set of 2000 from the available 19 456 samples is chosen at random and 20 estimates of the mean, using 100 samples for each, are calculated and plotted. 50% and 95% confidence intervals, obtained by using the mean and standard deviation calculated using all 2000 samples, are also shown. The latter values are given, along with the actual percentage of the 20 estimates that lie outside the confidence limits. In addition the number of the first sample of the whole block is shown—this can take any value between 0 and 17 456

(since there are 19456 samples altogether). The user can repeat this demonstration as often as he wishes by selecting the 'Repeat' action prompt. Alternatively, by selecting 'User tests' the user can choose his own parameters (confidence interval, number of samples per block, number of blocks, sampling increment and position of first sample) to obtain further confidence limit displays. In both these cases the user can opt to generate a screen dump file of the results, by selecting the appropriate action prompt. Figure 4.13 is an example of such dumps.

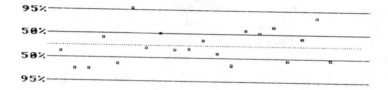

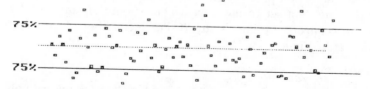

Figure 4.13.   Screen display examples of confidence limit charts obtained from the white noise data set.

The second option within this section is entitled 'Mean and mean-square value measurements', and can be chosen as an alternative to the confidence limit routines. This allows measurements of the means and standard deviations of blocks of samples to be obtained. The user is initially prompted to enter the various sampling parameters. As usual, default values are available and if these are selected the following results will be displayed for the Gaussian (white) noise data set:

Measurements from Gaussian (white) noise signal
8 blocks of 100 samples,
at intervals of 1 and starting at sample number 1
50% confidence limit has $z(0.25) = 0.674$

z (mean)	z (var.)	Mean	Std. Dev.
0.452	−1.118	0.310	36.235
−0.128	2.749	−1.980	46.539
0.849	−0.259	1.880	38.761
0.439	−0.775	0.260	37.263
−0.488	0.146	−3.400	39.897
1.204	0.190	3.280	40.018
−1.288	−0.847	−6.560	37.049
−1.040	−0.479	−5.580	38.129

In addition to the mean and standard deviation of each block the display includes the calculated values of the sampling parameter z. For the mean $z = (\hat{\bar{x}} - \bar{x})/(\sigma/\sqrt{N})$(see §4.3.1) and for the variance $z = (\hat{\sigma}^2/\sigma^2 - 1)/\sqrt{2/N}$ (see §4.3.2). In both cases the true values $\bar{x}$ and σ are calculated using all NB*NS samples—NS is the requested number of samples and NB the number of blocks. This option can therefore be used to produce cumulative distributions such as those in figures 4.7 and 4.9 or the more straightforward confidence limit charts such as figure 4.8.

At the conclusion of the display (and printout, if that was requested) the action prompts allow the user to do further calculations on the same data set (using 'repeat'), to choose an alternative data set or to change the printer status (using 'menu above'), or to return to the CHAP4 index.

4.6.3 Higher-order moments
Choosing this option from the CHAP4 index allows the user to obtain

estimates of higher-order moments, from the third to the sixth. After the usual title and window display, an initial prompt for the required moment is followed by a prompt asking whether the calculation of its value for the complete signal (all 19 456 samples) is required. After this calculation is completed (if requested), the user is then prompted for the usual set of sampling parameters to allow similar calculations for smaller blocks. All results are presented in normalised form, dividing the calculated $\bar{x^n}$ by the nth power of the standard deviation, obtained from the same data. A typical printer output is shown below. It should be noted that, if large sample sizes are requested the

$X\widehat{}4/(\text{sigma}\widehat{}4) = 2.961$ for all 19456 samples of the Gaussian (white) noise signal

12 blocks of 100 samples,
at intervals of 1 and starting at sample number 1

Values of $x\widehat{}n/\text{sigma}\widehat{}n$ are:

3.202	3.445	3.152	2.431	2.721	2.994	2.879	3.007
2.119	3.315	2.725	3.011				

calculation time becomes significant. This should not be surprising, since, particularly for the highest-order moments, the number of required multiplications becomes large (30 000 for the sixth-moment calculation for 10 000 samples). In the case of the BBC software, this is, of course, one of the reasons why the calculations are all performed in machine code; if this part of the program had been written in BASIC, the calculation time would be unacceptably long.

4.6.4 Probability density measurements The final measurements option in the CHAP4 index enables probability density functions to be obtained. Again, after the initial display the user is prompted for the sampling parameters—this time the number of 'slots', rather than the number of blocks, is required. The output includes the calculated mean and standard deviation (from the samples specified) and a screen plot of the estimated probability distribution. The data can also be printed if required (by initial setting of the printer status as usual) and the action prompts include an option to generate a screen dump file of the final plot. Note that the data printed out are in normalised form, i.e. the two columns given (see below) are for $\sigma p(x)$ and $(x - \bar{x})/\sigma$:

PROBABILITY DENSITY MEASUREMENTS
****

Using the Gaussian (white) noise signal

Number of samples (def.1000) 19456

starting at number (def.=1)

Sampling increment (def.=1)

Number of slots (def.=16)32

Mean = 6.522E−2   Sigma = 39.73

$(X-XB)/\text{Sigma}$	$\text{Sigma.p}(x)$
−3.223	6.637E−3
−3.022	5.871E−3
−2.821	8.679E−3
−2.619	1.302E−2
−2.418	2.68E−2
−2.217	4.442E−2
−2.015	5.616E−2
−1.814	8.883E−2
−1.612	0.1279
−1.411	0.1652
−1.21	0.2292
−1.008	0.2634
−0.8071	0.3027
−0.6057	0.3553
−0.4044	0.3645
−0.203	0.4184
−1.642E−3	0.3969
0.1997	0.3681
0.4011	0.3346
0.6024	0.3158
0.8038	0.2721
1.005	0.2154
1.206	0.1766
1.408	0.1279
1.609	0.1029
1.811	6.611E−2
2.012	4.237E−2
2.213	2.91E−2
2.415	1.94E−2

2.616	7.658E-3
2.817	5.871E-3
3.019	8.424E-3
3.22	0

where $\bar{x}$ and σ are the values calculated from the specified samples and given in the results. Further $p(x)$ estimates can be obtained and displayed by using the 'repeat' action prompt. Figure 4.14 shows a typical screen dump.

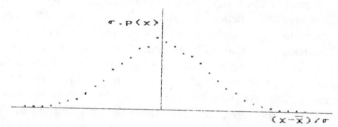

Probability distribution for the
Gaussian (white) noise   signal
19454 samples, starting at no.1
and with a sampling increment of 1
Mean = 6.369E-2   Sigma = 39.73

Filename PDF1

$\sigma.p(x)$

$(x-\bar{x})/\sigma$

Figure 4.14. Screen dump of a typical measured probability distribution of the white noise data set.

4.6.5 Final comments

All these routines can of course be run using any one of the available data sets. The user can therefore easily investigate the effects on estimate variabilities that arise from using correlated sample sets by, for example, loading the correlated (pink) noise data and using the mean value option to generate some confidence limit plots.

At any stage during the running of the options described above, the user can abort and return directly to the main CHAP4 menu by using the ESCAPE key.

⟩ Chapter 5

⟩ Digital Sampling Criteria: Time-Domain Statistics

⟩5.1 Introduction

Chapter 4 was devoted to consideration of the digital sampling requirements necessary when measurements of the amplitude probability density function, or any of its moments, are required. This chapter discusses similar questions but in the context of time-domain measurements—principally autocorrelation and spectral functions. In these cases, it should by now be clear that errors can arise not only from inadequate amplitude quantisation (see §4.2), from finite sample sizes or from sampling times too short to capture the low-frequency content of the signal but also from the quantisation in *time*. If the sampled values are separated too far apart in time, they could represent either low or high frequencies in the original signal. A trivial example is provided by considering the digitisation of a simple sinusoidal signal. If sampling occurred exactly once every cycle, a set of samples of identical amplitudes would be obtained, representing energy at zero frequency rather than the frequency of the original signal. This phenomenon is known as *aliasing* and will be considered in detail in §5.2. In later sections, we discuss the various sources of statistical error which can occur in measurements of autocorrelations (§5.3) and spectra (§5.4). It is assumed throughout this chapter that quantisation errors are negligible; this is nearly always the case in practice. There are specialised circumstances in which quantisation errors are deliberately introduced but these will not be discussed here.

It should also be emphasised at the outset that all the results for measurement variabilities quoted in the following sections assume a particular kind of signal—usually bandwidth-limited white noise. As in the case of amplitude-domain statistics, it is usually impossible to obtain analytic results for the expected errors in other cases. Since almost all signals of engineering significance are quite different from white noise, the formulae given here should be viewed only as estimates. They are, nevertheless, sufficient to provide some physical understanding of the nature of the errors and hence to give some guidance in the choice of suitable sampling parameters in real cases.

⟩5.2 Aliasing

Consider the digitisation of a continuous signal using a sampling rate $1/\Delta t$, so that individual samples are Δt apart in time. The maximum frequency which can be unambiguously recovered from the sample values is $1/(2\,\Delta t)$. Any energy in the original signal at frequencies higher than this will be 'folded back' and will appear as lower-frequency components in the energy spectrum. This arises essentially because of the circular nature of the Fourier transform process (equation (3.18)); recall that $\cos(2\pi f \Delta t) = \cos[2\pi(f \pm n/\Delta t)\,\Delta t]$, where n is an integer. Putting $f_a = 1/(2\,\Delta t)$, all data at frequencies of $f \pm 2nf_a$ have the same cosine function as data at frequency f and will therefore be indistinguishable from the latter. (The simple case of a sinusoidal signal sampled once every cycle, used in §5.1, is just a trivial example of this.) Consequently, measurements of the signal's energy content at frequencies f less than f_a will be contaminated by all energy at frequencies $2nf_a \pm f$. f_a is termed the aliasing frequency.

It is important to recognise that aliasing does *not* affect the measurement of the autocorrelation function of a signal if this is determined simply by performing the digital analogue of the relation $R(\tau) = \int_0^\infty x(t)x(t+\tau)\,dt$. Only when spectral data are required (and obtained either by cosine transforming $R(\tau)$ or Fourier transforming $x(t)$) does aliasing become relevant. The usual methods of surmounting the aliasing problem are either to low-pass filter the signal prior to sampling or to choose a sufficiently high sampling rate to ensure that energy levels in the signal above the aliasing frequency are negligible.

As an example of the effects of aliasing, consider pink noise having unit variance and an autocorrelation function $R(\tau) = \exp(-\alpha\tau)$. The energy spectral density function for the continuous signal is given by

$$E(f) = 4\alpha/(\alpha^2 + 4\pi^2 f^2)$$

with $E(0) = 4/\alpha$. For convenience, we define a normalised spectral density by $\hat{E}(f) = E(f)/E(0)$. Assume now that exact values of the autocorrelation function are available at intervals of $\Delta\tau$. With the usual expression for the digital equivalent of the cosine transform relationship, i.e.

$$E(f) = 2\Delta\tau\left(2\sum_{n=0}^{N-1} R(n\,\Delta\tau)\cos(2\pi fn\,\Delta\tau) - R(0) - R(N-1)\right)$$

(see §5.4), it can be shown that for the above pink noise signal the value of the normalised spectral density at the aliasing frequency $f_a = 1/(2\,\Delta\tau)$, is given by

$$\hat{E}_d(f_a) = \{[\exp(\alpha\,\Delta\tau) - 1]/[\exp(\alpha\,\Delta\tau) + 1]\}^2$$

provided that N is sufficiently large. The suffix d indicates that this is the digitally estimated spectral density. With $\alpha = 1$ for convenience, the ratio of the estimated to the exact spectral density is then approximately $\pi^2/4$, for small enough $\Delta\tau(\leq 0.5$, say). Typically, therefore, the time-domain quantisation leads to a measured energy level (near the aliasing frequency) about 2.5 times as large as the true energy in the original continuous signal.

Figure 5.1 is a plot of the true energy spectrum for this case, compared with what would be obtained digitally using an infinite number of samples (so that statistical errors are zero). Results for various sampling rates ($\Delta\tau$ values of 0.05, 0.1, 0.2 and 0.5) are shown. Two obvious points regarding figure 5.1 are worth emphasising. Firstly, as expected, the frequencies at which the digitally obtained spectra are inadequate rise as the sampling rate $1/\Delta\tau$ rises and it is evident that sampling rates corresponding to a $\Delta\tau$ of at most 0.1 are required to obtain adequate spectral data over an energy range of two decades. The implication of this for the choice of suitable sampling parameters in real cases is discussed in the following sections.

Secondly, note that spectral data at all frequencies above about $f_a/3$ are noticeably distorted by aliasing. The frequency at which such distortion becomes significant will depend on the shape of the original spectrum, however. If the rate of energy decay is much greater than the $1/f^2$ appropriate to pink noise, aliasing effects will be correspon-

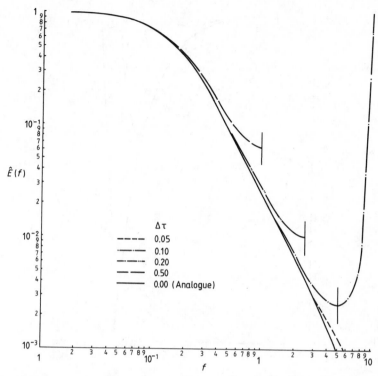

Figure 5.1. Ideal energy spectra obtained from digital transforms of the autocorrelation function of pink noise ($R(\tau) = \exp(-\tau)$), compared with the true spectrum: — — —, $\Delta\tau = 0.05$; — · —, $\Delta\tau = 0.1$; — · · —, $\Delta\tau = 0.2$; — —, $\Delta\tau = 0.5$; ———, $\Delta\tau = 0.0$ (analogue). The vertical lines denote the aliasing frequency.

dingly less noticeable, and vice versa. This is illustrated in figure 5.2, which shows true and estimated spectra for a signal having an autocorrelation given by $R(\tau) = 1/(1 + \tau^2)$. Such a signal has an exponentially decaying energy spectrum so that, as can be seen from figure 5.2, aliasing effects are much less significant than they are for pink noise, given an equivalent sampling rate. (Note that this signal is effectively the 'Fourier inverse' of pink noise—its spectrum is like the autocorrelation of pink noise, and vice versa.)

However, many physical processes yield signals having spectral characteristics not too dissimilar from pink noise; so the $f_a/3$ point provides a useful general rule. Unless the signal is suitably filtered prior to digitisation, one can generally expect distorted spectral data

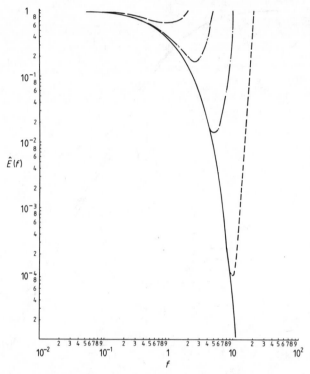

Figure 5.2. Ideal energy spectra obtained from digital transforms of $R(\tau) = 1/(1 + \tau^2)$. The curves have the same meanings as in figure 5.1.

at frequencies higher than about one sixth of the digital sampling frequency.

The software package provides a few set demonstrations of the effects of aliasing on estimation of the spectrum for the case of pink noise and the reader can also generate a wide variety of examples of his or her own, as described in §5.7.

⟩5.3   Autocorrelation estimation

Because of the exact Fourier transform relationship between the autocorrelation and the energy spectral density functions, there are essentially two ways of obtaining either. In the case of the measurement of $R(\tau)$, either one can deduce it directly from the signal by using the defining relationship (equation (3.11)) or it can be obtained by first Fourier transforming the signal to obtain the energy spectrum and then cosine transforming the result. It is, in many practical cases, more efficient (less computationally intensive) to use this latter route, but we begin by considering the conceptually simpler technique based on the definition of $R(\tau)$:

$$R(\tau) = \lim_{T \to \infty} \left(\frac{1}{T} \int_0^T x(t)x(t + \tau) \, \mathrm{d}t \right).$$

5.3.1 Direct estimates

On the assumption for the moment that N samples of $x(t)$ have been obtained at periodic intervals of $\Delta\tau$, an obvious digital version of this definition is

$$\hat{R}(m \, \Delta\tau) = \frac{1}{N - m} \sum_{i=0}^{N-m} x(i \, \Delta\tau)x(i \, \Delta\tau + m \, \Delta\tau)$$

$$m = 0, 1, 2, \ldots, M - 1. \quad (5.1)$$

Note that, although for the mth lag, there are $N - m + 1$ cross products, the divisor is just $N - m$; this gives an unbiased estimate. The more 'obvious' divisor, $N - m + 1$, would yield biased estimates, as discussed in the context of mean-square values in §4.3.2.

There are at most N consecutive estimates of $R(\tau)$, from $\hat{R}(0)$ to $\hat{R}((N - 1) \, \Delta\tau)$ inclusive. However, the number of cross products contributing to each estimate falls as m increases so that unless the maximum number M of m were chosen to be much less than the number N of samples, the variability of the estimates for long time lags would be significantly higher than for short lags. For this reason, it is usual to make $M \ll N$. This restriction is equivalent, in the analogue context, to requiring that the total signal sampling time be long enough to ensure adequate averaging of the longest lag components of $R(\tau)$.

In some circumstances, particularly if there is no intention to obtain spectral estimates from the autocorrelation values, it may not be necessary to calculate all the possible $R(\tau)$ values. Restricting the values of m to, say, the sequence 0, 2, 4, 6, ..., M (for integral $M/2$), or an even sparser subset, might be quite adequate. For example, the intention may merely be to obtain some overall measure of the dominant time scale in the signal. A common measure of the long-time-scale processes in $x(t)$ is given by

$$T_x = \int_0^\infty R(\tau)\, d\tau.$$

(In the case of pink noise with $R(\tau) = \exp(-\tau/T)$, T_x is precisely the time constant T.) It would generally then only be necessary to obtain a number of $R(\tau)$ estimates sufficient to define the complete curve from $\tau = 0$ to the value at which $R(\tau)$ has decayed to zero. The question then arises, 'Is it advantageous to sample $x(t)$ more rapidly than the minimum rate necessary to obtain $R(\tau)$ estimates at the required intervals?' Whilst sampling at a faster rate will certainly enable more cross products to be used for a particular $R(\tau)$ estimate, the individual samples—and hence the consecutive cross products—may be highly correlated, in which case the effect on the statistical accuracy may be minimal. We pursue this by reference to some theoretical results.

It can be shown that the variance of autocorrelation estimates in the general case is given approximately by

$$\text{var}[\hat{R}_n(m)] = \frac{1}{N-m} \sum_{i=-\infty}^{\infty} [R^2(i) + R(i+m)R(i-m)$$

$$- 4R(m)R(i)R(i-m) + 2R^2(i)R^2(m)] \qquad (5.2)$$

(Bartlett 1946a,b, Kendall 1973). Here, i and m should be understood as meaning $i\,\Delta\tau$ and $m\,\Delta\tau$, respectively, and $\hat{R}_n(m)$ is a normalised autocorrelation estimate defined by

$$\hat{R}_n(m) = \frac{1}{N-m} \sum_{n=0}^{N-m} [x(n)x(n+m)] \Big/ \frac{1}{N} \sum_{n=0}^{N} [x^2(n)]$$

$$m = 0, 1, 2, \ldots, M. \qquad (5.3)$$

The result assumes, amongst other things, that $x(t)$ is Gaussian. Now consider the typical case of a signal having an exponentially decaying autocorrelation (e.g. pink noise) given by $R(\tau) = \exp(-\tau/T)$ and assume that a sampling rate of $1/\Delta\tau$ is used so that autocorrelation estimates are obtainable at lag intervals of $\Delta\tau$; $R(i)$ corresponds to $\exp(-i\,\Delta\tau/T)$. Substitution into equation (5.2) then yields, after some algebra,

$$\text{var}[\hat{R}_n(m\,\Delta\tau)] = [1/(N-m)][\{[1 + \exp(-2\lambda)]/[1 - \exp(-2\lambda)]\}$$

$$\times [1 - \exp(-2m\lambda)] - 2m \exp(-2m\lambda)] \qquad (5.4)$$

where $\lambda = \Delta\tau/T$. Consider first the case of small λ, implying that consecutive samples from $x(t)$ are well correlated. In these circumstances it can readily be shown from equation (5.4) that reducing both the sampling rate and the number of samples (but keeping the total sampling time $N\Delta\tau$ constant) makes virtually no difference to the variability of the autocorrelation estimates at equivalent lag times. This is illustrated in figure 5.3, which shows the standard error $\sqrt{[\text{var}(\hat{R})]}$ obtained from equation (5.4) for a total sampling time of $50T$. The results obtained using just 100 samples of $x(t)$ with $\lambda = \Delta\tau/T = 0.5$ are very close to those with $N = 1000$ and $\lambda = 0.05$. In the former case, consecutive samples have an average correlation of about 0.67; increasing the sampling rate by an order of magnitude increases this to about 0.95 and it is evident that this order-of-magnitude increase in the number of available cross products for each $R(\tau)$ estimate is quite unjustified in terms of the resulting increase in accuracy. Of course, it would also normally increase the computational time by an order of magnitude. The experimental data also shown in the figure are discussed later.

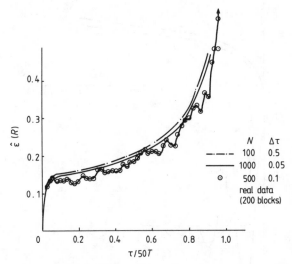

Figure 5.3. Standard error of autocorrelation estimates, as a function of lag: — · —, the Bartlett formula, $N = 100$, $\Delta\tau = 0.5$; ——, the Bartlett formula, $N = 1000$, $\Delta\tau = 0.05$; $\bigcirc$, real data, $N = 500$, $\Delta\tau = 0.1$ (200 blocks).

It turns out that the theoretical variability results are not too dependent on the form of $R(\tau)$. For example, if a signal having an autocorrelation of the form

$$R(\tau) = \cos(2\pi f\tau)\exp(-\tau)$$

is sampled in a way similar to the above, equation (5.2) can be used to find the variation in standard error shown in figure 5.4(a). For this calculation, $N = 100$ and the error in the fiftieth $R(m)$ contribution is plotted as a function of the period $1/f$ of the underlying periodicity in the signal. The standard errors (for the first 50 lags) are also shown

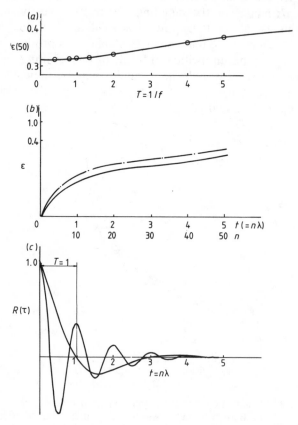

Figure 5.4. (a) Standard error in the fiftieth autocorrelation value ($m = 50$) of a signal having $R(\tau) = \cos(2\pi f\tau)\exp(-\tau)$ (see (c)). The results were obtained using equation (5.2) with $N = 100$ and $\Delta t = 0.1$ and are plotted as a function of the period $1/f$. (b) Standard errors for the first 50 lags, in cases where $1/f = 1$ (——) and $1/f = 4$ (—·—). (c) The autocorrelation curves for $1/f = 1$ and $1/f = 4$.

for the particular cases of $1/f = 1$ and $1/f = 4$ (figure 5.4(b)), together with the exact $R(\tau)$ variations corresponding to these (figure 5.4(c)). Clearly the dependence on the period is relatively weak, suggesting that the results of Bartlett particularised for pink noise can be used to give reasonable estimates of the variability errors for a wide range of signal types.

Consider now the case of large λ. This is essentially equivalent to a set of samples corresponding to white noise, since consecutive samples have a very low correlation. The variance of the correlation estimates then reduces essentially to $2/N$, as anticipated on the basis of statistical theory for uncorrelated samples (see §4.3.2). In practice, this limit is rather less relevant, since one normally only measures autocorrelation functions of signals which are known to have some definite time structure − $R(\tau) \neq 0$ for $\tau \geqslant 0$.

In view of the somewhat approximate nature of the arguments leading to equation (5.2), one might question the practical usefulness of this result. It is straightforward and instructive to test its accuracy by using simulated data sets. This can be done by the reader, by making use of the data supplied with the accompanying software. However, we first show some results obtained with a substantially larger data set. 100 000 samples representing pink noise with $R(m\,\Delta\tau/T) = \exp(-m\,\Delta\tau/T)$ and $\lambda = \Delta\tau/T = 0.1$ were generated and subsequently analysed in 200 blocks of 500 consecutive samples. Each block therefore had the same $N\,\Delta\tau/T$ (50) as used for the theoretical error curves in figure 5.3. All the possible autocorrelation estimates were obtained for each block and averaged over all NB blocks in order to yield the variability, defined by

$$\mathrm{var}[\hat{R}(m)] = \frac{1}{NB}\sum_{i=1}^{NB} [R_i(m) - \exp(-0.1m)]^2$$

$$m = 0, 1, 2, \ldots, M - 1.$$

Note, incidentally, that the mean value of $x(t)$ for each block was removed before forming the $R(m)$ estimates using equation (5.3). Figure 5.3 includes the resulting standard errors and it is clear that the Bartlett formula yields a satisfactory estimate of the variability. As anticipated, the errors rise rapidly as m approaches M because in this case $M = N$. It was pointed out earlier that it is normal in signal-processing applications to make $M \ll N$, also ensuring that M remains sufficiently large to allow $R(M\,\Delta\tau)$ to be close to zero (assuming that there are no strong periodicities in $x(t)$).

Some typical results obtained from the supplied data sets are shown in figure 5.5. Here, the pink noise data set (10 000 samples) was analysed in 50 blocks of 200 samples and 100 blocks of 100 samples, with just the first 50 autocorrelation estimates obtained. The results are compared with the Bartlett curves. Again, reasonable agreement is obtained, but it is noticeable that in all these cases the calculated errors are somewhat lower than the theoretical values. This is almost certainly due largely to the negative bias errors which can be shown theoretically to accompany $R(m)$ estimates obtained using equation (5.3) (see e.g. Kendall 1973).

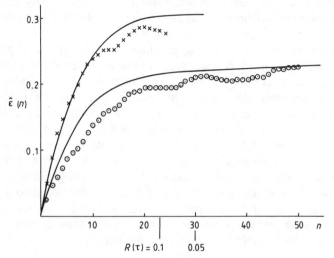

Figure 5.5. Measured standard errors of first 50 autocorrelation values estimated from 10 000 samples of a pink noise signal (from the simulated data available in the software): ×, 100 blocks of 100 samples; O, 50 blocks of 200 samples; ——, the Bartlett results.

It is important to emphasise that the consecutive autocorrelation estimates are *not* independent. In fact, they can be highly correlated so that quite smooth, but totally erroneous, autocorrelation curves can be obtained if insufficient samples are used. This is illustrated in figure 5.6, where three examples of the first 50 estimates obtained from the individual blocks of 200 samples are shown. The data represent just three of the 50 blocks used to obtain the variability estimates shown in figure 5.5 and it is clear that, although each set might look plausible in the absence of any prior information about $x(t)$, they are generally far from accurate representations of the true autocorrelation function.

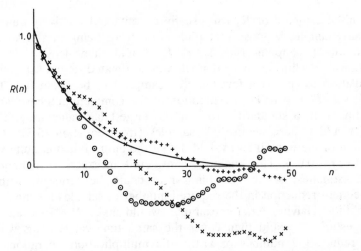

Figure 5.6.   Three examples of the first 50 calculated autocorrelation values, each from a different set of 200 samples of the simulated pink noise: ——, exact result, which is closely followed if all 19 456 samples are used.

It is possible to use the Bartlett result to derive a simple criterion for the required number of samples. On the assumption that the autocorrelation is similar to that of pink noise and that we require the last (Mth) lag to correspond to a time at which $R(\tau)$ has largely decayed to zero, the variability of this Mth estimate is given (from equation (5.4)) approximately by

$$\varepsilon^2 = 1/(N\,\Delta\tau/T)$$

where ε is the standard error and N is the total number of samples, assumed to be much greater than M. In fact, it can be shown that for signals containing strong periodic components the variability is always lower; so this result represents an upper bound. The required number of samples is then just

$$N = 1/(\varepsilon^2\,\Delta\tau/T).$$

Usually the effective time constant T of the signal is not known a

priori , but it can be approximately deduced using the first autocorrelation estimate $(R_1 = R(\Delta\tau/T))$ by $\Delta\tau/T = 1 - R_1$, so that the
required number of samples can be written

$$N = 1/\varepsilon^2(1 - R_1). \tag{5.5}$$

An initial estimate of R_1 can usually be obtained relatively quickly.
As an example, suppose that pink noise were being sampled using
some initial sampling rate $1/\Delta\tau$. R_1 would then depend on the
unknown T, but assume that it was estimated (from an initial
calculation using just a few hundred samples) to be about 0.9. This
implies $\Delta\tau/T = 0.1$. Then estimation of the complete autocorrelation
function with a standard error for the largest lag value lower than
0.05 would require about 4000 samples (from equation (5.5)). Calculation of the first 50 lags would define the entire function quite well
$(\exp(-5) < 0.01)$. This might be considered an excessive number of
$R(\tau)$ estimates; so the sampling rate could be reduced, with a
consequent reduction in the required number of samples for the same
variability. Halving $\Delta\tau$, for example, would make $N = 2000$ and 25
$R(\tau)$ estimates would then cover the same time range. This would
also reduce the number of arithmetic multiplication operations by
about a factor of 4.

Although only a relatively simple case has been considered it
should be emphasised that equation (5.5) is an upper bound on the
likely number of samples required for a given variability. It therefore
represents a convenient general criterion for determining the required
sampling parameters when only the general shape of the autocorrelation is required, although it should be recalled that the Bartlett result
assumed that $x(t)$ has Gaussian amplitude statistics.

We conclude therefore that, in many practical cases, rapid sampling
of $x(t)$ is not necessary if only autocorrelation information is required. It is sufficient merely to ensure that the sampling rate allows
autocorrelation estimates to be obtained at, say 20–30 consecutive
lags with the final lag time long enough to make $R(\tau)$ close to zero.
Of course, if the signal contains any dominant periodicities, so that
the autocorrelation function would be expected to oscillate about
zero, rather more lags may be required for an adequate representation of the signal. We emphasise again that, if spectral information is
required, sampling rates generally need to be considerably higher and
the autocorrelation estimates are best obtained via the spectra. This is
discussed in §5.3.2.

5.3.2 Autocorrelation estimates via spectra

As discussed in §3.2.2, the autocorrelation function can be written in terms of the spectral density function:

$$R(\tau) = \int_0^\infty E(f)\cos(2\pi f\tau)\ df$$

where, in terms of the original signal $x(t)$, $E(f)$ can be obtained as

$$E(f) = \lim_{T\to\infty} [|C(f)|^2/T]$$

with

$$C(f) = \int_0^\infty x(t)\exp(-2\pi ift)\ dt.$$

It is worth pointing out that the Fourier integral defining $C(f)$ only exists if $\int_0^\infty x(t)\,dt \leqslant \infty$, which is not true for a stationary random record $x(t)$, which theoretically extends over all time. However, one can only actually measure $x(t)$ over some finite time T, so that $C(f)$ is estimated by computing the finite Fourier transform:

$$C(f,T) = \int_0^T x(t)\exp(-2\pi ift)\ dt$$

which always exists.

In the digital calculation of $C(f)$, f (as well as $x(t)$) is restricted to take on discrete values and the usual digital equivalent of the finite Fourier transform is

$$C(k\,\Delta f) = \Delta t \sum_{n=0}^{N-1} x(n)\exp\!\left(\frac{-2\pi ikn}{N}\right) \qquad k = 0, 1, 2, \ldots, N-1$$

$$(5.6)$$

where $x(0)$, $x(1)$, $\ldots$, $x(N-1)$ are the digitised samples of the continuous signal $x(t)$. It is important to note that, unlike the case of a Fourier transform of a continuous signal, the transform of the discrete-time signal $x(n)$ is necessarily a periodic function of the frequency f since, for integral values of m,

$$C(2\pi f + 2\pi m) = \sum_{n=0}^{\infty} x(n)\exp(-2\pi ifn - 2\pi imn) = C(2\pi f).$$

This periodicity is present for exactly the same reason as the aliasing phenomena discussed in §5.2.

Equation (5.6) is usually called the discrete Fourier transform (DFT). Sampling $x(t)$ N times at intervals of Δt gives a sampling period of $T = N\Delta t$ and imposes a Nyquist (aliasing) frequency at $1/(2\Delta t)$ (see §5.2). Results are normally obtained with $\Delta f = 1/T$, which is the fundamental frequency—the lowest for which spectral information can be obtained from the finite set of N. Note that, in most practical cases, $x(n)$ is a real-valued sequence and values of $C(k\Delta f)$ beyond $k = N/2$ can be obtained from earlier values, since the Nyquist frequency is $N\Delta f/2$. We discuss appropriate implementation of the DFT for spectral measurements in §5.4.

To obtain autocorrelation estimates, two further steps are necessary. First, the values of k for the spectral density function must be obtained via

$$E(k\Delta f) = \text{Re}^2[C(k\Delta f)] + \text{Im}^2[(C(k\Delta f)]$$

where the two terms on the right-hand side are the squares of the real and imaginary parts of $C(k\Delta f)$, produced by the earlier DFT. Secondly, a further DFT must be performed, this time on the values of $E(k)$ rather than on the values of $x(n)$, and the real part of the results used to obtain $R'(m\Delta t)$:

$$\hat{R}'(m\Delta t) = \frac{1}{N\Delta t}\,\text{Re}\!\left[\sum_{k=0}^{N-1} E(k\Delta f)\exp\!\left(\frac{-2\pi imk}{N}\right)\right]$$

$$m = 0, 1, 2, \ldots, N-1.$$

(Recall that $\Delta f = 1/(N\Delta t)$.) However, these $\hat{R}'(m\Delta t)$ values will not be identical to the $\hat{R}(m\Delta t)$ values that would be obtained using the direct autocorrelation function definition (equation (5.1)). This is partly due to the circular effect in the calculation procedure and also to the fact that $1/N \neq 1/(N-m)$. The usual procedure to avoid the first effect is to perform the DFTs on number sets twice as long as the original with, for the first DFT on the basic data, the elements $x(N+1)\ldots x(2N)$ in the second half set equal to zero. The first half of the resulting $\hat{R}'(m\Delta t)$ values, when multiplied by $N/(N-m)$ to take account of the second effect, then have values identical with those that would be obtained from equation (5.1), using the same initial data set.

There are many computationally efficient algorithms for calculating the DFT, most of them based on one kind or another of the fast Fourier transform (FFT) technique first introduced in the 1960s. These usually accept a sequence of N complex numbers—$x(0)\ldots$ $x(N-1)$, where N is an integral power of 2—and yield N complex values of $X(n)$, calculated from

$$X(k) = \sum_{n=0}^{N-1} x(n) \exp\left(\frac{-2\pi ikn}{N}\right) \qquad k = 0, 1, 2, \ldots, N-1.$$

Abbreviating this expression by $X_k = \text{DFT}(x_n)$, we can list the steps required to obtain the autocorrelation function via the spectral density function as follows.

(1) Sample $x(t)$ N times at intervals of Δt and put

$$w_n = \begin{cases} x(n) \\ 0 \end{cases} \text{for} \begin{cases} n = 0, 1, 2, \ldots, N-1 \\ n = N, N+1, N+2, \ldots, 2N-1. \end{cases}$$

(2) Form $C_n = \text{DFT}(w_n)$ for $n = 0, 1, 2, \ldots, 2N-1$ and obtain $E_n = |C_n|^2$.

(3) Form $\hat{R}(m\,\Delta t) = [N/(N-m)]\,\text{Re}[\text{DFT}(E_n)]$, $m = 0, 1, 2, \ldots$, $N-1$, with the DFT performed with $n = 0, 1, 2, \ldots, 2N-1$.

This procedure will yield values of $\hat{R}(m\,\Delta t)$ identical with those obtained from equation (5.1), repeated here for convenience:

$$\hat{R}(m\,\Delta t) = \frac{1}{N-m} \sum_{n=0}^{N-m} x(n)x(n+m) \qquad m = 0, 1, 2, \ldots, M-1$$

(5.1)

with $M = N$ here.

The above process might seem to be somewhat tortuous compared with the relatively simple calculation using equation (5.1). However, its usefulness lies in the fact that FFTs are computationally very efficient; so, even if spectral information is not required, it can sometimes be quickest to use the spectral method to obtain autocorrelation estimates. Consideration of equation (5.1) shows that, for a calculation of all estimates ($M = N$), about N^2 arithmetic multiplication operations are required. It can be shown that an N-point FFT requires about $N \log_2 N$ multiplications, so that steps (1)–(3) above require about $4N \log_2(2N)$ which, for large N, is smaller than N^2. As an example of comparative timings, figure 5.7 shows the time required to calculate $R(\tau)$ using both methods. These were obtained from algorithms coded in FORTRAN 77, using a standard FFT and implemented on a 32-bit microprocessor (the Acorn Cambridge Workstation). Clearly, if $N \geqslant 128$, the spectral route is quicker and the difference in time exceeds an order of magnitude once $N \geqslant 2048$ (recall that FFTs generally require N to be an integral power of 2).

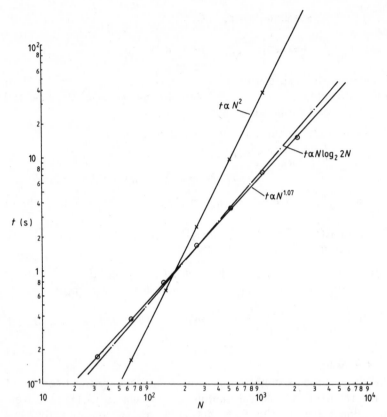

Figure 5.7. Timing for autocorrelation calculations via spectra (⊙) or directly from raw data (×). Calculations were performed in FORTRAN 77 on the 32-bit Acorn Workstation. Note that the N-point autocorrelation obtained via spectra used two $2N$-point FFTS.

Now, if spectral estimates are not required, the number of autocorrelation estimates needed will generally be low, as pointed out in §5.3.1, and probably much less than 128. In practice, therefore, if *only* $R(\tau)$ estimates are required, it is often quicker and certainly easier to use the more direct route to obtain them. Since the two methods, as outlined above, are statistically equivalent, the variabilities of the autocorrelation estimates obtained via spectra are identical to those discussed in §5.3.1. It is largely for this reason that, in the software package, only the direct method of estimating the autocorrelation function has been implemented. The reader can nonetheless use the available routines to illustrate the major points discussed above, as described in §5.7.2. We turn now to a closer consideration of the estimation of the spectral density function.

⟩5.4 Estimation of the spectral density function

The reader is first reminded of the relationship between the energy spectral density function $E(f)$ and the signal $x(t)$ or the autocorrelation function $R(t)$, which were first introduced in §3.2.2 and restated in §5.3. Since, in nearly all signal-processing applications the signals under scrutiny are not truly determinate, the relationship between the spectrum and the Fourier transform of the original signal is now reiterated in its general form (see §3.2.2):

$$E(f) = \lim_{T \to \infty} \left\{ (1/T) \operatorname*{ave}_{(j)} \left[|C_j(f)|^2 \right] \right\}$$

where the averaging process (representing the fundamental expectation operator) is carried out over j independent blocks of data. In digital form, this can be written

$$\hat{E}(k \, \Delta f) = \frac{2}{n_d T} \left(\sum_{j=1}^{n_d} |C_j(k \, \Delta f)|^2 \right) (\Delta t)^2 \qquad k = 1, 2, \ldots, N_b$$

with

(5.7)

$$C(k \, \Delta f) = \text{DFT}[x(n)] \qquad k = 1, 2, \ldots, N_b$$

as usual. There are n_d blocks of data, each containing N_b samples at intervals of Δt, so the sample time for each block is $T = N_b \, \Delta t$. The alternative method of obtaining $E(f)$ is to use the autocorrelation function:

$$\hat{E}(k \, \Delta f) = 4 \operatorname{Re} \{ \text{DFT}[R(k \, \Delta t)] \} \, \Delta t. \tag{5.8}$$

In both these cases the DFT is defined as in §5.3.2:

$$\text{DFT}[y(k)] = \sum_{n=0}^{N_b - 1} y(n) \exp\left(\frac{-2\pi i k n}{N_b} \right) \qquad k = 0, 1, 2, \ldots, N_b - 1$$

(5.9)

and $\Delta f = 1/(N_b \, \Delta t)$. Note that we use N_b here and reserve N for the total number of samples used ($N = n_d N_b$).

Now the DFT as defined above really amounts to using discrete values of $y(n)$ together with a simple trapezoidal rule to perform the integration. Without corrections, the results will therefore lie on a 'pedestal' of $[y(0) + y(N_b - 1)]/2$ and, in many applications, it is best to halve the first and last values of $x(n)$ (or $R(n)$) to remove this effect. Note also that any DC component will theoretically appear as an addition to the first spectral estimate $E(0)$, so the mean value of the N_b $x(n)$ samples should be removed before performing the DFT for each block. This should be done *prior* to halving the first and last samples; if it is not done at all, the pedestal correction will lead to distortion of the *complete* spectrum unless the mean value happens to be zero. An illustration of the pedestal effects is given in figure 5.8. 32 768 samples of pink noise were generated and the first 64 values of the autocorrelation function estimated in the direct way. $E(f)$ was then deduced using equation (5.8). In this case the variance of the complete set of samples was 0.10, so the spectral values obtained without halving $R(0)$ are distorted by the constant addition of $4R(0)/2 = 0.2$, which is increasingly significant as the frequency rises, as the figure demonstrates.

One of the more remarkable features of the spectrum is the fact that the (statistical) accuracy of an estimate of $E(f)$ at any particular frequency does not decrease as N_b increases, in complete contrast with the behaviour of $R(t)$ estimates. Furthermore, provided only that the basic N_b spectral estimates are calculated, these are virtually independent of each other if $x(t)$ is Gaussian (with the obvious proviso that the second $N_b/2$ estimates are a repeat of the first $N_b/2$ if $x(t)$ is real). This again is quite unlike the consecutive $R(t)$ estimates which, as discussed in §5.3.1, can be highly correlated.

In the context of analogue measurements, the variance of spectral density function estimates is proportional to $1/B_e T$, where B_e is the spectral bandwidth used for the measurements and T is the total measurement time. In the present context, B_e is equivalent to $1/(N_b \Delta t)$ (N_b samples per block) and $T = N_b n_d$ (n_d blocks), so the variance is proportional to $1/n_d$. In fact, it can be shown that the variance of the spectral estimates obtained using equation (5.7) is given approximately by

$$\text{var}[\hat{E}(f)] = \varepsilon^2 = \sigma^2/n_d. \tag{5.10}$$

This should be compared with the variance of $E(f)$ estimates obtained using equation (5.8), which, like that of the $R(t)$ estimates, will be proportional to $1/N_b n_d$, provided that the number m of $R(t)$ estimates is restricted so that $m \ll N$. The difference is illustrated in

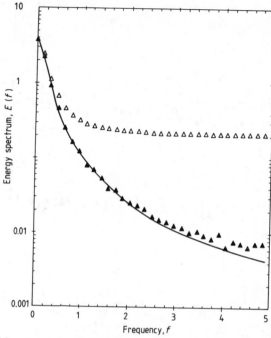

Figure 5.8.   Spectral estimates from 32 768 samples of simulated pink noise, obtained via 64 $R(t)$ estimates: $\triangle$, using the true $R(0)$; $\blacktriangle$, with $R(0)$ halved; full curve, exact spectrum. Calculations performed on the Acorn Workstation.

figures 5.9 and 5.10. 32 768 samples of white noise were simulated and then analysed to obtain spectral estimates in both ways. Firstly, all samples were used to obtain 64 consecutive $R(t)$ estimates via equation (5.1), with subsequent calculation of the spectrum via equation (5.8). Secondly, the sample set was split into n_d blocks of N_b samples, with $n_d = 32$ and $N_b = 1024$ (or $n_d = 512$ and $N_b = 64$) and the spectrum calculated via equation (5.7). The spectral estimates, normalised so that the integral under $E(f)$ (up to the aliasing frequency) was unity, are plotted against frequency in figure 5.9. For the second case, only those estimates at frequencies equivalent to those of the 32 estimates arising from the first method are shown, for clarity. It is immediately apparent that (with $n_d = 32$) the variability of the estimates obtained by transforming the original data is considerably greater than for those obtained via the autocorrelation. Of course, if the spectrum were obtained via equation (5.8) having

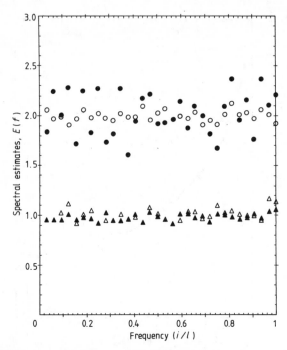

Figure 5.9. Spectral estimates from 32 768 samples of simulated white noise obtained via $R(t)$ ($\bigcirc$, $\triangle$) or directly from $x(t)$ ($\bullet$, $\blacktriangle$) where 32 blocks of 1024 samples ($\bullet$) or 512 blocks of 64 samples ($\blacktriangle$) were used: $\bigcirc$, $\bullet$, $E(f)$ incremented by unity for clarity.

calculated *all* the possible $R(t)$ estimates from the complete data set ($N_b n_d$ of them), then the same variability would be expected from both methods. The difference seen in figure 5.9 arises essentially because only a relatively small number of $R(t)$ estimates were used, minimising the effects of relatively larger variabilities in the $R(t)$ estimates at the longest lag times: $m = O(N)$. The resulting set of $E(f)$ values is sometimes referred to as the truncated periodogram, indicating that it was derived from a truncated set of autocorrelation estimates.

Further data sets of the same size were generated and similar calculations performed with n_d equal to 16, 64, 128, 256 and 2048 and corresponding values of N_b (such that $n_d N_b = 32\,768$). The average variance of the 32 spectral estimates was computed in each case and is shown in figure 5.10. Agreement with the theoretical

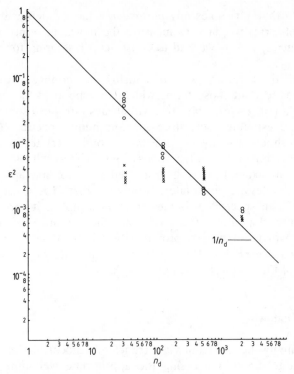

Figure 5.10. Variability of spectral estimates as a function of the number of independent blocks used: ×, estimates obtained via the truncated autocorrelation; ○, estimates obtained directly from the raw data. In each case, a total of 32 768 samples were used.

variance $(\varepsilon^2 = 1/n_d)$ is reasonable, bearing in mind the limited number of estimates (32) used to obtain the results. The variability of the spectral estimates obtained via the autocorrelation, for the identical data sets, is included in figure 5.10.

The unsatisfactory nature of spectral estimates obtained from just one block of data $(n_d = 1)$, however long the block, arises essentially because we are trying to estimate a *continuous* function rather than a single parameter (or set of parameters). The property of 'smoothness' which we would like the set of estimates to possess is a concept not normally met in classical estimation theory. Similar difficulties were apparent in the estimation of the (ideally) continuous probability density function. The power at a *particular* frequency, like the probability that $x(t)$ takes any *particular* value, is identically zero; as in physical terms we have to measure the power in some small band of frequencies, just as we had to construct a histogram for estimating $p(x)$.

As implied above, one way of reducing the variance of the spectral estimates is to omit those terms which correspond to the 'tail' of the autocorrelation. In principle, this will always introduce some bias into the spectral estimates but, since for continuous spectra $R(t) \to 0$ as $t \to \infty$, it should usually be possible to omit $R(t)$ terms at large t without this effect becoming too serious. The results in figure 5.9 constitute an example of this truncated periodogram approach and further examples are given later. If only the first M $R(t)$ estimates are used to obtain spectral estimates from N samples, it can be shown theoretically that both the bias and the variance of the $E(f)$ estimates will tend to zero provided that both $M \to \infty$ and $N \to \infty$ but in such a way that $M/N \to 0$. There are many ways to achieve this: one of the simplest is to put $M = \sqrt{N}$.

$\rangle$5.5   Windowing

Now obtaining the spectrum via a truncated autocorrelation is directly equivalent to smoothing (using the appropriate weighting function) the 'raw' periodogram that would be obtained by transforming *all* the original data (i.e. equation (5.7) with $n_d = 1$). The weighting function is normally called the spectral window, whereas the effective weighting function used to truncate the autocorrelation is termed the lag window. In the example used above, the lag window was of a simple rectangular form, defined by

$$\lambda(k) = \begin{cases} 1 & k \le M \\ 0 & k > M. \end{cases}$$

The estimated spectral density function (equation (5.8)) could then be written as

$$\hat{E}(k\,\Delta f) = 4\,\mathrm{Re}\{\mathrm{DFT}[\lambda(k)R(k\,\Delta t)]\}\,\Delta t \qquad (5.11)$$

which is equivalent to

$$\hat{E}(k\,\Delta f) = (2\,\Delta t^2/T)|\mathrm{DFT}[W(k)x(k)]|^2$$

where $W(k)$, the spectral window, can be shown to be the (discrete)

Fourier transform of the lag window $\lambda(k)$. One of the most commonly used lag windows—often known as the Bartlett estimate—is that defined by

$$\lambda(k) = \begin{cases} 1 - k/M & k \leqslant M \\ 0 & k > M \end{cases}$$

which can be shown to correspond to a spectral window having a decaying $\sin^2 \theta$ function. The point to note here is that it can be demonstrated theoretically that smoothed spectral estimates obtained by applying this spectral window to the raw periodogram obtained from all the data values (or, equivalently, obtaining the spectrum by transforming all the autocorrelation estimates weighted with the above lag window) is essentially identical with dividing the original record up into a smaller number of blocks and averaging the periodogram at fixed frequencies over all blocks. The distinct advantage of the latter technique is that the amount of data storage required at each stage in the computation is considerably smaller. In the present context of the analysis of continuous signals, this approach is generally quite satisfactory, but it should be recognised that, in other contexts (usually those in which a more limited number of data samples are available), windowing techniques assume increasing importance. There is a considerable literature on the whole subject of windowing, as a method of smoothing spectral estimates, and the reader should consult the more detailed texts if further information is required (the book by Priestly (1981) contains an extensive discussion). Before turning to a brief discussion of simple criteria for choosing the various parameters for spectral estimates, however, there is one further point about simple (rectangular) truncation of the autocorrelation that should be emphasised.

The rectangular lag window defined above is less satisfactory than the Bartlett estimate if the estimated autocorrelation function has not decayed close to zero for values of k near M. It is easy to see why this is so, for the Fourier transform of $\lambda(k)$ for the rectangular window is essentially a $(\sin \theta)/\theta$ function (the Dirichlet kernal) which has substantial negative lobes either side of its maximum value at $\theta = 0$. The spectral density estimates from equation (5.11) will therefore take negative values at certain frequencies and this is obviously not physically acceptable. As an example, consider a pink noise signal having an autocorrelation function defined as usual by $R(\tau) = \exp(-\tau/T)$ and assume that values of $R(\tau)$ for $\tau \leqslant pT$ are used to obtain the spectral density function via equation (5.11) ($\lambda(\tau) = 0$ for $\tau \geq pT$). The ideal spectral density function, i.e. what would be obtained if the $R(\tau)$ estimates were exact, is then given by

$$E(f) = 4 \int_0^{pT} \exp\left(-\frac{\tau}{T}\right) \cos(2\pi f\tau)\, d\tau$$

which, after some algebra, reduces to

$$E(f) = E_t(f)\{1 + [\alpha \sin(p\alpha) - \cos(p\alpha)]/\exp p\} \qquad (5.12)$$

Here $\alpha = 2\pi fT$ and $E_t(f)$ is the true spectral density function that would result if the complete autocorrelation function were available ($p = \infty$), i.e.

$$E_t(f) = 4T/(1 + 4\pi^2 f^2 T^2).$$

It is clear that, quite apart from the effects of aliasing (arising from finite Δt) or variability (arising from imperfect $R(\tau)$ estimates), the spectral density function will be unphysical, in that it will take negative values when $[\cos(p\alpha) - \alpha \sin(p\alpha)]/\exp p \geq 1$ and, in any case, oscillate about the true values with increasing amplitude as the frequency increases. This may not be a serious effect if p is large enough to ensure that the negative regions only exist (if at all) at frequencies where the true energy is very much less than its maximum value. However, if the higher-frequency part of the spectrum is a region of particular interest, truncation effects can be serious and so should be avoided. Figure 5.11 is an example of the effect. $\hat{E}(f)$ from equation (5.12) is shown for a case in which $p = 3.2$ (and $T = 1$, for convenience), compared with the true spectrum. Also shown are the results of the digital computation of the spectrum (via the autocorrelation) from a simulated data set of 32 768 values. In this latter case, $\Delta t = 0.025$ and the first 128 $R(\tau)$ estimates were computed, so that the last available estimate corresponded to a time lag of $128\,\Delta t = 3.2 = p$, as required. Although $R(128\,\Delta t)$ is only about 0.04, the truncation effect is clearly visible at the higher frequencies, with the measured spectrum falling well below the true spectrum. Note that negative values do not occur in this case and that, because of the particular value of Δf, only points near the minima of the oscillations are picked out by the spectral calculations; a different choice of $M\,\Delta t$ could equally yield spectral estimates near the maxima. Similar demonstrations can be obtained using the accompanying software, as described in §5.7.

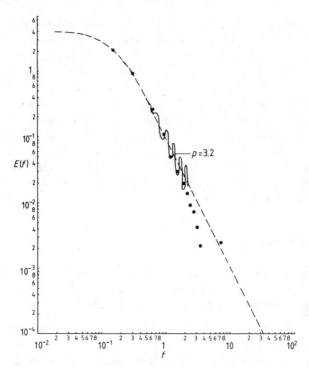

Figure 5.11. Truncation effects on spectral estimates derived from the autocorrelation function of pink noise: — — —, ideal spectrum; ———, spectrum obtained from truncated autocorrelation (equation 5.12) with $p = 3.2$); ●, digital estimates from a set of 32 768 samples, with $\Delta t = 0.025$ and using the first 128 calculated autocorrelation values only (so that $M\Delta t = 3.2$).

The reader should note that windowing techniques which seek to minimise or remove entirely these unphysical negative energies—such as the Bartlett window mentioned earlier—are in one important respect not very sensible. They all have the effect of destroying, to some degree at least, what starts as perfectly good data, or $R(\tau)$ estimates. If $0 \leqslant \lambda(\tau) \leqslant 1$ for $\tau \geqslant 0$, which is generally the case for lag windows other than the simple rectangular one, the 'good' estimates of $R(\tau)$ are changed prior to being used to obtain the spectral estimates (via equation (5.11)). Intuitively, this seems silly. Techniques have been developed to avoid this unwanted degradation of $R(\tau)$ values; most of these are based on principles of 'maximum entropy' (Burg 1967, Lacoss 1971). The essential idea is to use the information already present in the existing $R(\tau)$ estimates to extend the autocorrelation function beyond the $k = M$ point. The additional $R(\tau)$ estimates have a variability somewhat greater than they would have if deduced from the original set of $x(t)$ values, but it has been conclusively demonstrated that spectral estimates obtained by transforming the extended sequence of autocorrelation values can be much more satisfactory than those obtained either by simple truncation (using only the original $R(\tau)$ values) or by using any of the standard forms of lag window (see, e.g., Fagih 1980). This is a rather specialised topic and, since it is certainly not a standard technique, will not be discussed further.

⟩5.6   Summary and examples

It should by now be clear that digital estimation of the spectral density function, from a continuous stationary signal, requires specific decisions concerning, at the very minimum, the following:

(1) the sampling *rate*, i.e. the Δt increment at which the signal is to be digitised:

(2) the number N_b of samples to be used in each block if spectra are to be deduced directly from the data or, equivalently,

(3) the maximum lag value M if spectra are to be deduced via the autocorrelation function;

(4) the total number N of samples to be used.

The choice of Δt is usually fixed by aliasing considerations as discussed in §5.2. Since N_b (or M) determines the increments in frequency at which the (uncorrelated) spectral estimates will be available, its choice is generally governed by the required resolution in the complete estimated spectral density function. This is often called the 'resolvability' requirement. Clearly, if there are likely to be one or more narrow peaks in the spectrum, $N_b \Delta t$ must be sufficiently large to ensure that $\Delta f = 1/(N_b \Delta t)$ is small enough to resolve the peaks satisfactorily. The problem is similar to that discussed in the context of the measurement of the probabability density function, where the amplitude 'slot' width needs to be small enough to resolve any humps in $p(x)$. Since the variance of the final spectral estimates is $O(N_b/N)$, making N_b large will, for a given required variance in the spectral estimates, inevitably make N large. If there are no serious limits on the record length that can be obtained, it is therefore usual to choose N_b first, according to the required resolvability, and then to make N sufficiently large to ensure an acceptable level of variability in the spectral estimates.

Now, in order to resolve spectral peaks adequately, one usually needs some idea of the bandwidth of the signal. Spectral bandwidth is an important physical notion and it is common engineering practice to define it as the distance between the 'half-power' points—the points on either side of the spectral peak at which the energy has dropped to one half of its maximum value. Very often, of course, one has only a vague idea *prior* to a particular measurement of the likely bandwidth but it is often helpful to recall that generally the spectral bandwidth will be small if the autocorrelation function decays slowly, whereas it will be large if $R(\tau)$ decays quickly. Examples of the two extremes are provided by a constant DC signal (or a pure sine wave) and white noise, respectively.

In practice, one often knows whether the signal contains dominant periodicities, which could imply a small bandwidth, or is largely of a wide band type whose dominant time scale is typified by the rate of decay of the autocorrelation function. Given some estimate of the bandwidth, one would then require, typically, four or five spectral points within it to resolve the peak adequately. As an illustration, consider the case of a signal having an autocorrelation defined by $R(\tau) = \cos(2\pi\tau)\exp(-\tau)$. This signal has a single dominant frequency of 1 Hz (with τ measured in seconds for convenience). It was simulated using 131 072 samples at time intervals Δt of 0.01. Note that this implies an aliasing frequency of 50 Hz, so that the example assumes that frequencies above this value are of no interest. Three representations of the signal are shown in figure 5.12. Whilst evidently the periodicity is not always very clear, a rough estimate of its frequency could be made simply by inspection of the signal.

On the basis that the periodicity is dominant enough to be visible, we start by assuming a spectral bandwidth of, say, 0.5 Hz, which means that the required $\Delta f = 1/(N_b \Delta t) = 0.1$. With $\Delta t = 0.01$, N_b must be, say, 1024 and this would give 512 spectral estimates up to the aliasing frequency. The normalised mean-square percentage error of the spectral estimates is (equation (5.10))

$$\mathrm{var}[\hat{E}(f)]/\sigma^2 = 1/n_d.$$

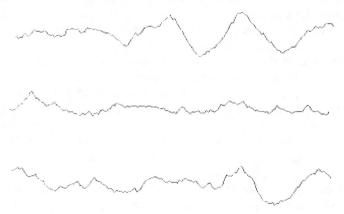

```
This is a time history of the
   ´ Another second order   ´ signal

(i.e. Data file :2.$.DAT5H8)

Saved on file S1
```

Figure 5.12.   Screen displays of typical time histories of a second-order signal.

So about 100 blocks of data would be required to keep the variance of the errors below 1%. This then implies a total sample number $N_b n_d$ of over 100 000, corresponding to a signal time of 1000 s. This may seem an excessive time, but recall that we are using a convenient ideal signal which has a dominant frequency component at 1 Hz. In many practical cases the important frequencies will be much higher; so shorter sample times would suffice to define the spectrum adequately, although, of course, low-frequency components can only be properly averaged with relatively long sampling times. Figure 5.13 shows the spectrum resulting from a direct calculation (using equation (5.7)) with 128 blocks of 1024 samples in each. Note that the aliasing effect can be seen at frequencies beyond about 20 Hz but, for many purposes, this spectrum would be more than adequate. The results cover an energy range of about four decades and are generally very close to the ideal spectrum, which is also shown in the figure.

It is often helpful to obtain an initial crude set of spectral estimates, in order to obtain some 'feel' for the general spectral shape. In this case, for example, we could have started by using rather fewer samples per block (hence reducing the 'resolvability' by reducing Δf) and, in addition, accepting a greater variability by reducing the number of blocks. Figure 5.13 includes results obtained by using just the first 64 blocks of 128 samples each; this requires only one sixteenth of the earlier sampling time, but it yields results which for some purposes would be quite satisfactory.

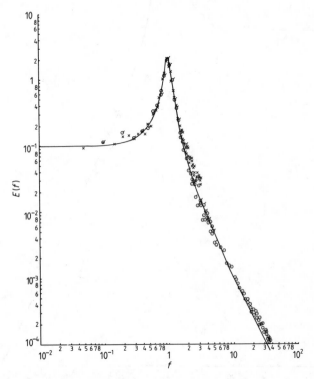

Figure 5.13. Spectra calculated directly from 131 072 samples of a simulated second-order signal having an ideal $R(\tau) = \cos(2\pi\tau)\exp(-\tau)$: 128 blocks of 1024 samples with $\Delta t = 0.01$, so that the aliasing frequency is 50 (O) or $\Delta t = 0.04$ (O′, not all shown). ×, 64 blocks of 128 samples with $\Delta t = 0.16$. Full curve is exact spectrum.

A final example is shown in figure 5.14, obtained from one of the data sets available in the accompanying software package. This is a simulation of a signal having $R(\tau) = \cos(\pi\tau)\exp(-\tau)$; samples every $\Delta t = 0.025$ are available, but the results have been obtained (via a

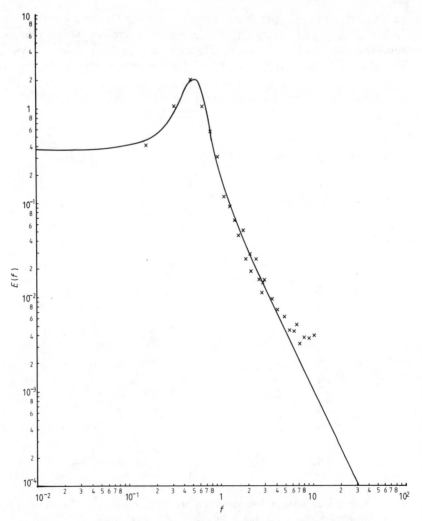

Figure 5.14. A spectrum calculated using the available software package, for the second-order signal having $R(\tau) = \cos(\pi\tau)\exp(-\tau)$. The results were obtained via the first 128 calculated autocorrelation values, with $\Delta t = 0.05$, so that the aliasing frequency is 10. Full curve is exact spectrum.

128-point autocorrelation) using only every second sample, i.e. $\Delta t = 0.05$. Again, for many purposes, this spectrum would be adequate. Other examples can be generated by the reader, as explained

in the following sections, in order to explore the various effects of
changing the sampling rate, the number of samples per block and the
number of blocks. The options also allow spectral estimates to be
made via autocorrelation estimates, although this can be rather a
lengthy process.

We conclude this section by emphasising that, in practice, it is
often necessary to use considerable trial and error before finding the
most suitable combination of parameters in a particular case. What
would be acceptable in one situation may be quite inappropriate in
another. Different physical phenomena can obviously display a very
wide range of spectral content and there are really no detailed criteria
that will enable a user to obtain adequate spectral information if he
or she has neither any prior notion about the signal content nor any
feeling about the limitations of digital spectral analysis.

〉5.7 Computer exercises

Demonstrations and exercises appropriate to most of the material in
the preceding sections can be undertaken by running CHAP5 from
the main index. The general protocol, screen layouts and action
prompts are similar to those used in the CHAP4 routines; whilst the
user could in principle use CHAP5 without prior knowledge of the
CHAP4 facilities, it is anticipated that most readers would normally
only progress to CHAP5 after at least some assimilation of the
CHAP4 material.

Selection of CHAP5 leads to a display of a further index, referred
to later and in the action prompts as "CHAP5 index". It is again
suggested that the user selects each of the three program options in
turn and studies the various facilities available by using the simple
default values for all the parameters, before attempting experiments
of his or her own. The following sections describe the various
options, but some initial comments about the structure of this part of
the software are appropriate here.

First, as noted in §5.3.2, the algorithm used for autocorrelation
estimation is that based on the direct method embodied by equation
(5.1). Machine code routines perform all the time-consuming parts of
the calculation (note that for calculation of a 64-point autocorrelation
from 16 000 samples about 1 million multiplications are required).

Secondly, spectral estimation can be performed either using the previously calculated autocorrelation (using equation (5.8)) or directly via the raw data (using equations (5.7)). Although block sizes cannot exceed 256 in the present software, this will give the user some feel for the relative merits of each, in terms of estimate variabilities and calculation time, as discussed in §5.4.

Thirdly, it should be emphasised that, in the context of some applications of signal processing, 20 000 samples does not represent a very large sample size. As indicated in §5.6, accurate measurements of spectral quantities would often require a considerably larger set of data. However, 20 000 *is* large enough to allow perfectly adequate illustrations of the theoretical ideas presented earlier and, in some circumstances, would be quite sufficient for real measurements.

Next, note that quantitative presentation of the calculated autocorrelation and spectral functions of the simulated data requires some assumptions about the effective sampling rate of the basic data sets. As discussed in Appendix A, the pink noise signal (having $R(\tau) = \exp(-\tau)$) was generated as a set of data samples spaced apart in time by 0.1 units whereas the second-order signal (having $R(\tau) = \cos(\pi\tau)\exp(-\tau)$) used a 0.025 sample increment. The random (uniform $p(x)$) and Gaussian noise signals did not of course require a particular Δt for their generation, but the software assumes a 0.1 sample interval between each of the 19 456 data values in all the quantitative data presentations in these sections of the software package.

Finally, note also that the spectral calculations all make use of a FFT routine (implementing equation (5.9)) written originally in assembler and kindly supplied by Structured Software. (The copyright for this part of the software is held by them.) It would be exceedingly difficult for users to extract this part of the present software for insertion into their own without some knowledge of its basic structure, so users who wish to make individual use of this routine should contact the suppliers directly. Their address is given in the Acknowledgements.

5.7.1 *Aliasing demonstrations*

This routine does not use any of the supplied data sets. Instead, it uses a pre-recorded set of 256 consecutive $R(\tau)$ values which were calculated using $R(\tau) = \exp(-\tau)$, with τ ranging from 0 to 6.375. These values are held in the file named RTAU on the system disc;

they represent the *exact* autocorrelation function that would be measured using an infinite number of samples of a perfect pink noise signal, obtained at a sampling rate of 40 Hz (with $\Delta\tau = 0.025$, say). Spectral estimates obtained by Fourier transforming this set of data will therefore be exact and will only differ from the ideal analogue spectrum $4/(1 + 4\pi^2f^2)$ by virtue of the aliasing effects generated by a finite sampling rate.

After selecting this option, the autocorrelation values are loaded into memory and the user is then prompted to specify how many of them are to be used in the spectral estimation. If, say, only 32 are specified (the default number) then every eighth $R(\tau)$ value, starting at $R(0)$, will be used as the data to be transformed, so that the aliasing frequency will be $1/(16\Delta t) = 2.5$ Hz. After the subsequent calculation of the 32 spectral estimates, these are plotted on the screen. Logarithmic scales are used since this highlights the aliasing effect. Figure 5.15 is a screen dump of the default results; note that the second 16 estimates are simply a repeat of the first 16 in the reverse order, as explained in §5.4. The calculation can be repeated using a larger (or smaller) number of the available $R(\tau)$ values and the user will note that this leads to a higher (or lower) aliasing frequency with a correspondingly better (or worse) fit to the ideal analogue spectrum.

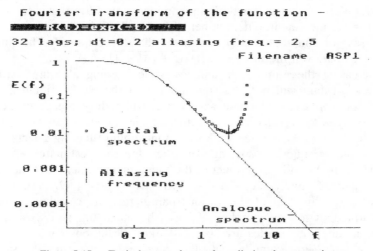

Figure 5.15.   Typical screen dump of an aliasing demonstration.

No printout is available in this option, but the user can obtain screen dumps if required. Return to the Chapter 5 index is effected in the usual way.

5.7.2 Autocorrelation estimation

This routine allows direct calculation of the autocorrelation function using any batch of data from any of the available data sets. The software essentially implements equation (5.1). Selection of the option from the Chapter 5 index will lead to the usual screen heading and window displays followed, after choice of the required data set and printer status, by prompts for the required number of samples, sampling increment, position of the first sample, lag increment and number of lags. Some explanation of these last two parameters is required. Assume that you have chosen 1024 samples (the default) with a sampling rate of 2 (the default), starting at the first available sample (the default). These are now the only data available with which to construct the autocorrelation estimates. If a lag increment of 1 is chosen (the default) the first available $R(\tau)$ estimate (not counting $R(0)$) will correspond to $\overline{x(t)x(t + 2\Delta\tau)}$, where $\Delta\tau$ is the effective time delay between each sample of the *original* set of 19 456 (recall that this is 0.025 for the second-order signal and 0.1 for the all the others). A lag increment of 2 would make the first $R(\tau)$ estimate that given by $\overline{x(t)x(t + 4\Delta\tau)}$. In general, the first estimate will be $\overline{x(t)x(t + ij\Delta\tau)}$, where i is the sampling increment (2 in this case) and j is the lag increment. The number of lags is then simply the number (M, say) of consecutive $R(\tau)$ estimates that are required. The final $R(\tau)$ estimate will therefore be $\overline{x(t)x(t + (M - 1)ij\Delta\tau)}$.

Increasing the number of samples while keeping all other parameters constant will not alter the number of the $R(\tau)$ estimates, or their location along the time axis, but it will reduce the theoretical variability in the results. Note also that, if spectral estimates are to be subsequently obtained (see §5.7.3), the maximum time lag effectively fixes the frequency spacing of the spectral estimates $\Delta f = 1/(M - 1)ij\Delta\tau$ and the product of the sampling increment and the lag increment effectively fixes the aliasing frequency $f_a = 1/(2ij\Delta\tau)$.

After specification of the various parameters the calculations are performed; if a large number of lags (the maximum is 128) and a large number of samples are specified, the calculations could take a few minutes. For example, a 64-lag autocorrelation from 8192 samples requires about 120 s of calculation time. Note that, if the user intends to obtain spectral estimates from the autocorrelation esti-mates, then it is sensible to choose the number of lags to be an integral power of 2. Once the calculations are complete, the results may be printed out if desired (by pre-setting of the printer status, as usual) and the screen plot can also be sent to a disc file for later dumping to a printer. Figure 5.16 shows typical screen dumps, obtained using the above example in the case of Gaussian white noise and pink noise. The usual action prompts after the display is complete allow the user to undertake similar calculations on the same or different data sets or return to the Chapter 5 index.

The user clearly has maximum flexibility in his choice of the necessary parameters, allowing a very wide range of experiments to be devised, as in the CHAP4 program.

5.7.3 Spectral density estimation
Selection of this option from the Chapter 5 index leads, after the usual choice of a data set and a printer status, to a further action prompt which allows the user to choose whether he wishes to obtain the spectrum from a previously calculated autocorrelation (via equa-tion (5.8)) or from the raw signal data (via equation (5.7)). In the former case, it must be emphasised that calculation of the autocor-relation should be undertaken *immediately* prior to entry—via the Chapter 5 index—of the spectral routine. If, after the autocorrelation calculation, the user returns to the main index (by using the BREAK key, or in any other way) before moving to the spectral routines, then it will not be possible to obtain the spectrum from that autocorrelation. If this is attempted, a warning is displayed.

On the assumption that this option has been successfully entered, the user is prompted for the required number of autocorrelation lag values to be used. The default value is the maximum possible, i.e. the total number of lag values that were previously calculated, but the user can specify a smaller number if required. If all available $R(\tau)$ estimates are used, then the aliasing frequency will be determined by the original lag increment, as explained in §5.7.2. If, however, a smaller number of the $R(\tau)$ estimates are used, then the aliasing frequency will be correspondingly reduced. Choosing only a quarter of the available estimates, for example, will mean that every fourth $R(\tau)$ value will be used.

After the spectral estimates have been obtained, the user is prompted for the screen plot in the usual way. The display includes

Calculated autocorrelation of

64 lags from 8192 samples
with sample increment of 0.1
and lag increment of 0.1

Filename AC1

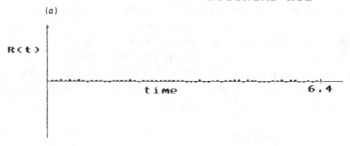

Calculated autocorrelation of

64 lags from 8192 samples
with sample increment of 0.1
and lag increment of 0.1

Filename AC2

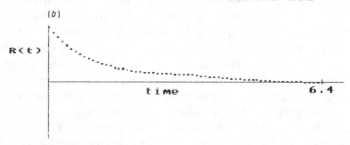

Figure 5.16. Screen dumps of typical autocorrelation calculations of (a) white noise and (b) pink noise.

information on the aliasing frequency, etc, and also gives the value of the total power, using the $R(0)$ data. By generating a printout of the results (in the usual way), the user can check that this total power is equal to the total area under the plot of $E(f)$ against f. Screen dumps can also be produced, and figure 5.17 shows typical ones, giving, for figure 5.17(a), results obtained from a 64-lag autocorrelation previously calculated using 8192 consecutive values of the Gaussian white noise signal (see figure 5.16(*a*) for these latter results). Note that in both this and the raw signal option only the first half of the spectral estimates are presented (32 values in this case); those beyond the aliasing frequency are ignored and the user is reminded that they are identical to the first half but in the reverse order. Note also that linear scales are used in the screen plots for both this and the alternative spectral option.

By responding in the usual way to the final action prompt the calculation can be repeated using the same or a different number of lags or the user may return to the menu.

Choice of the alternative spectral raw signal option allows calculations of spectral estimates from the basic simulated signals. After initiating this routine and specifying the required data set and printer status, the user is prompted for the sampling increment INC, the number NS of samples per block and the number NB of blocks. Apart from the necessary restrictions that the number of samples per block must be an integral power of 2 with a maximum of 256 and that INC*NB*NS must not exceed 16384, any values may be chosen for these various parameters. The FFT routine is called for each block of NS samples in turn, with the spectral density estimates being updated after each. During the calculations the display includes a continuous update of progress by showing the most recently completed block. As an example of the time required for the calculations, spectral estimates from 128 blocks of 64 samples requires about 165 s in the case of the BBC version of the software. Note that the FFT calls for all these 128 blocks require only about 40 s in total; a large proportion of the calculation time is taken up by the recovery of the individual estimates after each FFT and the updating of the running sums of these estimates. This is all programmed in BASIC; later versions of the software may include faster routines for this part of the process.

As usual, the spectral results can be printed out directly and/or a screen dump of the displayed spectrum can be produced. Figure 5.18 contains typical examples, showing the spectral estimates resulting from a calculation using 128 blocks of 64 samples of the Gaussian white noise and pink noise signals. These can be compared directly with the spectra shown in figure 5.17.

The usual responses to the final action prompt allow the user to repeat similar calculations on the same or different data or to return to the Chapter 5 index.

(a)

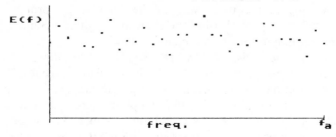

(b)

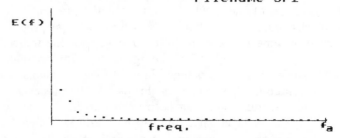

Figure 5.17.   Screen dumps of the spectral data calculated using the autocorrelations shown in figure 5.16. Note that the axes are linear.

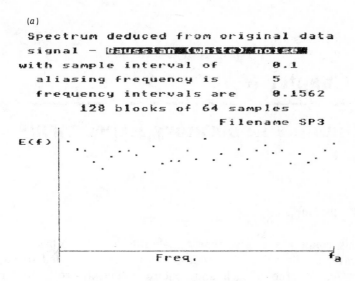

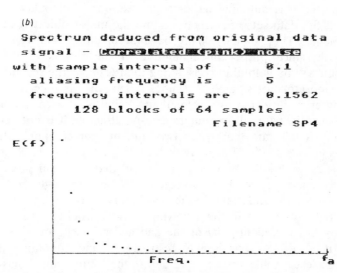

Figure 5.18.   Screen dumps of typical spectral calculations from the raw signal
for (*a*) white noise and (*b*) pink noise.

〉 Chapter 6

〉 Sample Laboratory Experiments

〉6.1   Introduction

In this chapter a typical pair of laboratory classes, based on the software, is described. These classes have been designed to require about one afternoon each and require a 'blackboard' introduction from the teacher, outlining the essential features of the software, the way that the data sets are structured and the major objectives of the experiments. They should ideally be undertaken in relatively quick succession, since proper understanding of the results of the first afternoon will be helpful in considering the results of the second.

If a laboratory of suitable machines (each with its own disc drive and printer facilities) is available, then obviously a large number of students can be catered for simultaneously, although it is not generally helpful to allocate more than two students per computer. In the author's classes at the University of Surrey, each pair of students has their own copy of the basic system disc so that they can proceed at their own pace. The students have generally been through the lecture course prior to undertaking the laboratory classes. In fact, the experiments described in the following sections would be very taxing for the average student if he or she had had no background; so they would need to be considerably restructured and simplified if the laboratory classes are to take place before the lecture course has been completed. Since the author's lectures usually include demonstrations using the software—mainly from CHAP2 and CHAP3 programs—the students at least have some visual familiarity with the package. There is then no reason why individual students should not undertake the experiments on their own and in their own time, particularly if they have a copy of this book. For the more interested student, this would probably lead to a better understanding of the material, because it would allow more time for simply 'playing' with the various signals and trying a wider range of measurements.

The whole of a laboratory sheet which might be handed to each student at the start of the first class is given in §6.2. This contains the major objectives of the experiments and much of the basic information required by the student in order to proceed. It assumes that the student has not read the accompanying textbook but it does *not* contain detailed instructions on how to run the software, because it is also assumed that the teacher will cover that verbally and be on hand when difficulties arise. It has also been found that the weaker students require rather more explanation of the structure of the data sets, particularly when faced with decisions concerning sampling rates and lag increments in the context of autocorrelation measurements.

It should be emphasised that many different kinds of experiment, with different major objectives, could be designed with this software. The following merely represents a relatively brief 'overview' type of experiment, but for more extensive courses it could readily be replaced by a longer sequence of laboratory classes, each concentrating on just one or two aspects of the material. Note that, in most of the author's lecture courses to date, the material covered in the text is not included in its entirety. This would require a considerably longer course (of 40–50 h plus laboratory time, typically). Note also that, since the lecturer can easily generate his or her own data sets for use with the software, including those obtained from a real transducer, there is considerable flexibility in the range and types of experiment that can be devised.

⟩6.2 Typical laboratory sheets

Final-year signal-processing laboratory classes

INTRODUCTION
The following two experiments are designed to enable the student to obtain a firmer grasp of the underlying principles governing the accurate measurement of various properties of fluctuating signals. They have been developed using a software package which accompanies the text of a recent book on the digital analysis of stationary data.

Algorithms to produce sequences of numeric data having the properties of various kinds of signal are now commonplace, as are microcomputers with substantial amounts of memory. The use of simulated signals (rather than real ones obtained from real transducers) to investigate digital sampling and measurement techniques is therefore straightforward. It is also an attractive concept since no hardware other than the basic computer system is required, and the user can make any 'measurement' that he or she likes—provided that he or she has the software tools available. These tools are provided in this software package; they allow measurement of both amplitude- and time-domain statistics of a range of simulated signals, using sampling parameters specified by the experimenter.

In the two experiments you will undertake in these classes, attention is concentrated on two of the basic questions affecting measurement accuracy:

(1) How many digital samples are required?
(2) How rapidly must the sampling be done?

The experiments are intended to demonstrate that the answers to these questions depend essentially on the kind of measurement being undertaken and on the nature of the signal itself.

In both experiments, use is made of the available simulated signals. These consist of 19 456 consecutive 1-byte (8-bit) integers. The data sets have been produced using standard algorithms which need not concern you now. They represent random noise (with a uniform probability density function), Gaussian (white) noise, correlated pink noise and a second-order signal having an autocorrelation which oscillates as it decays. The latter can be thought of as, for example, a noisy periodic signal. In every case the mean value of the entire data set is close to zero and the standard deviation is such that the signal virtually 'fills' the amplitude range (from -127 to 127 for 8-bit integers). It is loaded from disc into memory when required and can be accessed in various ways by the software to allow rapid calculation of means, mean squares, spectral density functions, etc.

For any measurement, you will be prompted for the required sampling parameters, such as the number of samples and the spacing between each sample (i.e. whether consecutive samples are required or whether only every nth sample is needed). The latter is equivalent to a 'sampling rate' if the original sequence is thought of as the basic signal. You may also specify the location of the first sample; remember that only 19 456 samples are available.

In Experiment 1, measurements of the probability distribution and its various moments are made whereas, in Experiment 2, attention is concentrated on measurements of the autocorrelation and spectral density functions. Different types of signal are used and you are asked, generally, to comment on the differences between your measurements and the theoretically expected results for a signal of the appropriate type. Each experimental programme consists of a set of instructions, each item of which has one or more associated questions. After completion of the second class, you should prepare a formal write-up of the two experiments, presenting the various results and answering the associated questions. You will find it helpful to begin to prepare the material for this after the first class and before the second class. This will help both to identify those measurements that you may wish to repeat and also to suggest supplementary tests that may be informative.

EXPERIMENT 1

The software should be loaded by inserting the disc into drive 0 and pressing the SHIFT and BREAK keys simultaneously, releasing the latter first. An initial index is displayed which gives the titles of the various chapters (which correspond to the chapters in the accompanying textbook). Selection of any one of these is achieved by using the cursor and RETURN keys, as indicated.

Begin by entering CHAP2 and trying each of the options in order to familiarise yourself with the different kinds of signals that are available. In this CHAP2 (and in CHAP3) the cursor keys are used for switching between options. This first experiment uses the simulated random noise, Gaussian (white) noise and the correlated (pink) noise signals; so move on to CHAP4 and display these signals on the screen in turn, using the appropriate option. Note that they are the files DAT0N8, DAT1N8 and DAT2N8, respectively, and are resident on side two of the disc.

(1) Using the appropriate CHAP4 options, measure the mean, the standard deviation and the next four higher-order moments (i.e. $\overline{x^n}$, $n = 3$–6) of the probability distribution for the random-noise data, using all available 19 456 samples.

Questions

(*a*) Given that the simulated signal lies in the amplitude range $-127 \leqslant x \leqslant 127$, discuss the difference between your measured standard deviation and the expected result.

(*b*) Show that the *n*th moment of a uniform probability function which is zero outside $|x| = a$ is given by

$$\overline{x^n}/\sigma^n = 3^{n/2}/(n + 1)$$

provided that *n* is even. Compare this result with the measured values, and discuss how you would expect the measured results to vary if you chose much smaller sample sizes.

(2) Repeat (1) for the Gaussian (white) noise data set.

Questions

(*a*) Compare the higher-order moments with the analytical values expected for Gaussian data ($\overline{x^4}/\sigma^4 = 3$, $\overline{x^6}/\sigma^6 = 15$). Discuss possible reasons for any differences and suggest ways in which the data could be arranged (i.e. stored, digitised or whatever) so that these differences might be less significant.

(*b*) What implications do these results have in the context of digitising analogue signals for later analysis on a computer?

(3) Using the CHAP4 mean-square option, obtain 10 independent estimates of the mean value of the Gaussian noise, using 1000 samples for each. Also obtain 20 independent estimates using only 100 samples for each.

Questions

(*a*) What is the meaning of a 90% confidence interval?

(*b*) Prepare confidence limit charts in which these mean-value estimates are plotted against estimate number, i.e. 1–20 for the 100-sample estimates. (You may use the confidence limit option within the mean-value option of CHAP4 to study the appearance of such charts—producing a screen dump on your printer if you wish— see the note below on screen dumps). Include the 50% and 90% confidence intervals on your charts, calculated by using the standard deviation measured using all 19 456 samples. Recall that the confidence interval is defined by $\sigma z_\alpha/\sqrt{N}$, with $z_\alpha = 0.674$ and 1.645 for the 50% and 90% intervals, respectively.

(*c*) Discuss the results in terms of your expectations deduced from the confidence intervals.

(4) Obtain in a similar manner 20 independent estimates of the mean value of the correlated (pink) noise signal using 100 samples. Use first *consecutive* samples (i.e. a sampling increment of one) and repeat using only every ninth sample.

Questions

(*a*) Prepare a further confidence limit chart for these results and explain why the two sets of mean-value estimates, obtained using different sampling rates, are so different. What would you expect if the same comparison had been made for the white noise signal?

(5) Obtain probability density estimates of the white noise signal using both 32 and 128 amplitude 'slots' and 1000 samples and 10 000 samples. Generate screen dump files of the results on 10 000 samples (see below) and send these to the printer. Obtain also a quantitative printout of the 32-slot results.

Questions

(*a*) With reference to these results, discuss the various ways of increasing the accuracy of probability density function estimates, emphasising any advantages or disadvantages of each.

(*b*) Plot the 32-slot results on your own graph and include the ideal Gaussian probability distribution function

$$\sigma p(x) = \exp[-(x - \bar{x})^2/2\sigma^2]/\sqrt{2\pi}.$$

Discuss the comparison between the two.

SCREEN DUMP FILES

After graphical presentation of results on the screen, the action prompts allow you to generate a disc file containing the screen display, for later dumping to a printer. If this option is selected, you are prompted for a drive number and a filename. Use the default value for the former and choose any appropriate name for the latter (it must not exceed seven characters). The file can subsequently be sent to the printer by selecting the PRNTDMP option from the main index and following the displayed instructions.

EXPERIMENT 2

The intention of this experiment is to give the student a firmer grasp
of the physical significance of the autocorrelation and the spectral
density functions, the relationship between them and some of the
problems which arise in their measurement. As in Experiment 1, just
three types of signal are considered: the Gaussian white noise, the
correlated noise and the second-order (noisy periodic) signal. Begin
by entering CHAP3.2 to remind you of some typical ideal signal
time-domain characteristics. Recall that signals with identical
amplitude-domain statistics can have very different spectral contents,
and vice versa.

Now move to CHAP4, select the display option and look at these
three signals in turn. Given a suitable time scale, how well could you
estimate the dominant frequency in the second-order signal?

(1) Select CHAP5 and using the default options obtain an estimate
of the first 16 lag values of the white noise signal from just 1024
samples. Generate a screen dump file and then (immediately) obtain
the corresponding spectral density estimates by returning to the
CHAP5 index and selecting the spectral density (from previous
autocorrelation) option. Generate a further screen dump file of these
results. (*Do not* return to the main index to dump your screen file of
the autocorrelation results to the printer before making this second
measurement.) Now send these two files to the printer.

Questions

(*a*) How do these results compare with those anticipated for ideal
white noise? Discuss the differences and explain how 'better' results
could be obtained.

(*b*) Is the level of accuracy in these results similar to that in the
probability density estimates of white noise obtained from 1000
samples in Experiment 1, and if not, why not?

(2) Select the aliasing demonstration option from the CHAP5
index. Choose to obtain spectral estimates using all 256 autocorrelation values (which are *exact* here) and also using just 32 of them.
You do not need to generate screen dump files, but make a note of
the general form of the results, and how they compare with the ideal
pink noise spectrum. Note also the value of the aliasing frequency in
each case.

Questions

(*a*) Remembering that the ideal autocorrelation values used to obtain these two sets of estimates are effectively from data sampled at time intervals given by 0.025 and 0.20 s, say, respectively, check that the displayed aliasing frequencies are correct.

(*b*) Explain what happens to the spectral estimates at frequencies above this aliasing frequency and discuss what determines how badly the estimates below it compare with the ideal spectrum.

(*c*) What is the relationship between the *time* intervals between the autocorrelation estimates and the *frequency* intervals between the spectral estimates deduced from them?

(3) The simulated pink noise signal has an ideal autocorrelation function given by $R(\tau) = \exp(-\tau)$, with consecutive data points separated in time by 0.1 s, say, for convenience. The ith autocorrelation value measured from an infinite number of such samples would therefore have the value $\exp(-0.1i)$, assuming every sample were used. (In the aliasing demonstration the $R(\tau)$ values are actually set to be *exactly* these ideal values.) Obtain estimates of the first 16 $R(\tau)$ values for the correlated (pink) noise signal, using just 2048 data points and printing out the results. Use a sampling interval of 1 (i.e. $\Delta t = 0.1$).

Question

(*a*) Plot these results as $R(\tau)$ against τ and compare them with the ideal $R(\tau)$ function. Why would these values not lead to satisfactory spectral estimates, even if they were made much more accurate by greatly increasing the number of samples?

(4) Obtain further autocorrelations from the pink noise signal with, first, a lag increment of 4 and 32 lags and, secondly, a lag increment of 1 and 128 lags; use a sampling increment of 1 and all available 19 456 samples in both cases. Note that these are relatively lengthy calculations (why?)—the second will take about 6 min. After each of these, proceed immediately to the spectral option and calculate the resulting spectra from the $R(\tau)$ values, printing out the results in both cases.

Question

(a) Plot the two sets of spectral estimates as $E(f)$ against f on log–log paper (preferably 3 × 3 decades) and include the ideal pink noise spectrum function given by $4\sigma^2/(1 + 4\pi^2 f^2)$, where σ^2 is the signal variance (you can deduce this from data included in the printout). Discuss these results with particular reference to the effects of the aliasing frequency and the finite number of samples used.

(b) How would the sampling parameters need to be changed if a higher spectral resolution were required (i.e. if Δf were to be lower) without any reduction in the aliasing frequency?

(c) Why could the aliasing frequency *not* be increased beyond 5 Hz in spectral estimates from this data set?

(5) Select the option that calculates spectral estimates directly from the original data set (the 'raw' signal). Choose the second-order signal, which ideally has $R(\tau) = \cos(\pi\tau) \exp(-\tau)$, and obtain estimates (without a printout) using a sampling increment of 4, and eight blocks of 256 samples each. Generate a screen dump file and send this to the printer.

Questions

(a) Estimate from this screen dump, in any way that you wish, the frequency of the dominant spectral peak. How close is this estimate to the ideal value implied by the above autocorrelation function?

(b) Suppose that the major point of interest was this dominant frequency f_0. How, in general terms, would you arrange the digital sampling from an original analogue signal having similar characteristics to enable a more accurate estimate of f_0 to be obtained in an efficient way? For an analogue signal having the same dominant frequency as in this present case, illustrate your answer by suggesting appropriate values for the digital sampling rate and also the number of samples per block and the number of blocks that you would use in the subsequent spectral analysis.

(c) In what other ways could you have estimated the dominant frequency in this simulated signal without doing a spectral calculation at all?

〉 Appendix A

〉 The Simulation of Random Signals

〉A.1   Introduction

This appendix outlines the methods used to generate the four data sequences available in the software package. Listings of the short algorithms (in BASIC) are included, so that the reader can easily generate his or her own data sequences having similar or different characteristics. The essential criterion for all the data, whether generated in this way or obtained directly by digitisation of a real analogue signal (using the BBC's on-board ADC, for example) is that it should be a sequence of signed 8-bit integers, stored continuously in memory from the address &3000 upwards. As supplied, all the software assumes that a maximum of 19 456 values are available; so, when loaded into memory, the data normally occupy from &3000 to &7BFF. If the user were to generate shorter sequences, it should be borne in mind that, unless appropriate changes are made to the BASIC routines in the software package, these will still assume that 19 456 values are available and only trap sampling parameters which imply a greater number. The location of these traps can be easily found by inspection of the CHAP4 and CHAP5 program listings.

After generation of a data sequence, it can be saved on disc with the usual operating system command

*SAVE filename 3000 7BFF

(assuming that there are 19 456 values). The reader is reminded that the four files supplied are named DATnN8, with $n = 0, 1, 2$ or 3 (see below). A simple approach to using additionally generated data files is therefore to save the new file on side 2 of (a copy of!) the system disc using one of these names; it will then automatically be loaded by the software whenever the equivalent (original) signal sequence is requested. There are two obvious disadvantages with this approach. Firstly, the original sequence will be overwritten and therefore unavailable on this disc. Secondly, the original data sequence descriptor (random noise, correlated noise, etc) may not be appropriate for the new signal. A rather more flexible facility has therefore been provided and, before discussion of the techniques used to generate the supplied data sequences, this will be described.

This facility allows the number of accessible data files to be increased to a maximum of 10, each having its own appropriate descriptor. *n* may therefore take additional values in the range 4–9 and, if these are used, the user *must* also update the DNAME file. The format of this file (which is on side 1 of the supplied system disc) is straightforward. It contains first a single integer denoting the number of available data sequences (N, say) followed by N strings each of 23 characters in length, forming suitable descriptors for the corresponding data sequences. A listing of the program used to generate this file for the supplied software is given below.

```
10 REM   PROGRAM FOR GENERATING FILE "Dname"
20
30 REM   containing NF signal data set descriptors
40
50 REM   ***********************************************
60
70 DIM data$(9)
80 PROCdata
90 DN$=":0.$.Dname"
100 SP$="                       "
110 OUT=OPENOUTDN$:NF=4:PRINT£OUT,NF
120
130 REM   Note that the maximum possible NF is 10
140
150 FOR I=1 TO NF:
    PRINT£OUT,LEFT$(data$(I−1)+SP$,23):NEXT
160 CLOSE£OUT
170
180 END
190
200 DEFPROCdata
```

210 data$(0)="   Random noise   "
220 data$(1)="Gaussian (white) noise   "
230 data$(2)="Correlated (pink) noise   "
240 data$(3)="Data with R=Cos(t)e(−t)"
250
260 REM   Add more file descriptor statements here
270 REM   if required; make sure NF in line 110
280 REM   is updated appropriately
290
300 ENDPROC

With appropriate modification this can be used to update
DNAME. Note that the values of n must be consecutive; if the user
wishes to add two extra data files to the system, these must have the
names DAT4N8 and DAT5N8, the variable NF in the NDATA file
must have the value 6, and the fifth and sixth 23-character strings
must contain the appropriate descriptors.

Once the additionally generated files have been saved on disc and
the DNAME file has been appropriately updated, all the available
data sequences can be operated on by all the routines in CHAP4 and
CHAP5. Additionally, they can be displayed graphically (as $x(t)$
against t) using the appropriate option available in CHAP4 (see
§4.6). Note that 10 files each of 19 456 samples virtually fills one side
of a standard disc, so the user may need a separate disc for any
screen dump files that he or she wishes to produce (these require disc
space of at least 19 456 bytes since they are mode 1 memory dumps).
There is room on side 1 of the system disc for one screen dump file.

⟩A.2   Generation of random noise (DAT0N8)

Most computer languages include algorithms, usually coded as intrin-
sic functions, for the generation of random sequences. BBC BASIC is
no exception and the function RND(I) can be used for this purpose.
With I = 1, repetitive calls of this function will yield a sequence of
numbers in the range (0, 1) having (ideally) a uniform probability
distribution. The only additional statements in the routine given
below are those required to scale the data so that they take values in
the range from −127 to 127. This implies a standard deviation of
about 73 (see A.4).

```
10 REM   ROUTINE TO CREATE A RANDOM
20 REM   (Uniform PDF) DATA SET
30
40 VDU22,7:HIMEM=&2FFF
50
60 INPUT "NUMBER OF SAMPLES", NS
70 U1=40*RND(1)
80
90 REM   Loop for data generation
100
110 FOR I=0 TO NS−1:U=254*(RND(1)−0.5)
120 ?(&3000+I)=INT(U+0.5)
130 NEXT
140
150 REM   Data stored in &3000 upwards (line 120)
```

〉A.3   Generation of Gaussian (white) noise (DAT1N8)

To obtain a random sequence having a Gaussian probability density function the method of Box and Muller (1958) was used. Let X_1 and X_2 be independent random variables with uniform probability distribution in the interval $(0, 1)$, i.e. as generated, say, in the way indicated above. Consider the random variables

$$W_1 = [-2 \ln(X_1)]^{1/2} \cos(2\pi X_2)$$
$$W_2 = [-2 \ln(X_1)]^{1/2} \sin(2\pi X_2).$$

Box and Muller showed that W_1 and W_2 form a pair of independent random variables with a Gaussian probability distribution and with a zero mean and unit variance. A sequence of random variables W_i may therefore be generated using the equations

$$\begin{aligned} W_{i-1} &= [-2 \ln (X_{i-1})]^{1/2} \cos(2\pi X_i) \\ W_i &= [-2 \ln (X_{i-1})]^{1/2} \sin(2\pi X_i) \end{aligned} \tag{A.1}$$

and the program listed below implements this. Note that again the generated values are scaled so that they lie in the range from -127 to 127 and have a standard deviation of around 40. The implications of this latter value (being about one third of the maximum range) for calculation of the higher-order moments is discussed in §4.2.

```
10 REM   ROUTINE TO CREATE A GAUSSIAN
20 REM        (WHITE) NOISE DATA SET
30
40 VDU22,7:HIMEM=&2FFF
50
60 INPUT "NUMBER OF SAMPLES",NS
70 U1=RND(1):P2=2*PI:E=40*SQR(2)
80
90 REM   Recursive loop – equation A1 in Appendix
100
110 FOR I=0 TO NS STEP2:U2=RND(1)
120 F=E*SQR(-LN(U1)):D1=INT(F*COS(P2*U2)+0.5):
    D2=INT(F*SIN(P2*U2)+0.5)
130 IFABS(D1)>127THEND1=127*SGN(D1)
140 IFABS(D2)>127THEND2=127*SGN(D2)
150 ?(&3000+I)=D1:?(&3001+I)=D2
160 U1=U2:NEXT
170
180 REM   Data stored in &3000 upwards (line 150)
```

⟩A.4   Generation of correlated (pink) noise (DAT2N8)

The requirement here was for a signal having Gaussian amplitude-domain statistics but an autocorrelation function of the form $R(\tau) = \exp(-\tau)$. Such a signal may be simulated by passing white noise through a first-order linear filter, which is simply the physical realisation of a linear differential equation or, in the discrete case (as required here), a linear difference equation. Many texts describe linear systems and their input–output characteristics; here we state only the basic differential equation which embodies the required connection between the output $x(t)$ and the input (white noise) $W(t)$ together with the recurrence relations required to simulate it. For pink noise of the above form, the filter transfer function is

$$\dot{x}(t) + ax(t) = \sqrt{2a}\, W(t)$$

and the corresponding recurrence equation is

$$x(i + 1) = \exp(-a\,\Delta t)\, x(i) + \sqrt{2/a}\,[1 - \exp(-a\,\Delta t)]W(i) \quad (A.2)$$

where $W(i)$ is the discrete Gaussian (white) noise and an arbitrary value can be taken for $x(0)$. DAT2N8 is a sequence produced using the program listed below, with $a = 1$ and $\Delta t = 0.1$. Consecutive data points are therefore separated in time by 0.1 units. The software package uses this value for the basic sample increment, as discussed in the previous sections. If an effectively more rapidly sampled simulated signal is required, then it is only necessary to change the a factor in the above equation. This amounts to a simple scaling of the time axis and obviates the need to change the assumed $\Delta t = 0.1$ in the software package.

For example, putting $a = 0.1$ and keeping $\Delta t = 0.1$ would effectively increase the sampling rate by a factor of 10, so that the first possible autocorrelation value (other than $R(0)$) would theoretically be $\exp(-0.01)$. Of course, the effective total sampling *time* would also be reduced by a factor of 10 (assuming that the same total number of samples were used) so that spectral estimates of the lower-frequency components of the data sequence would inevitably be less accurate. If the user *does* choose to produce a sequence with a different Δt, it would be necessary to change the appropriate data printout statements in the CHAP5 routines to ensure consistency in the printout of the results.

```
10 REM   ROUTINE TO GENERATE FIRST-ORDER DATA
20 REM             SET — with R(t)=exp(−at)
30
40 REM   DT is the required time increment between samples
50 REM   NS is the required number of samples
60
70 VDU22,7:HIMEM=&2FFF
80 INPUT "a,DT,NS",A,DT,NS
90 REM   Set-up various parameters
100
110 P1I=2*PI:AE=EXP(−A*DT):B=SQR(2/A)*(1−AE)
120 X0=0:U1=RND(1)
130 REM   Recursive loop — equation A2 in Appendix
140
150 FOR I=0 TO NS−1 STEP2:U2=RND(1)
160 F=SQR(−2*LN(U1)):W1=F*COS(P1I*U2):W2=F*SIN(P1I*U2)
170 X1=AE*X0+B*W1:X0=AE*X1+B*W2
180 Y1=130*X1:IFABS(Y1)>127THENY1=127*SGN(Y1)
190 Y2=130*X0:IFABS(Y2)>127THENY2=127*SGN(Y2)
200 ?(&3000+I)=INT(Y1+0.5):?(&3001+I)=INT(Y2+0.5)
210 U1=U2:NEXT
220
230 REM   Data stored in &3000 upwards (line 200)
```

Note again the presence of an amplitude scaling factor in the program above, chosen to give a standard deviation of about 40 for the resulting sequence. Some trial and error may be necessary before the required standard deviation can be obtained; this process can be considerably helped by using the probability option in CHAP4 to check that both the resulting mean and standard deviation and the overall probability distribution are satisfactory.

〉A.5   Generation of a second-order process (DAT3N8)

In this case, the requirement was for a signal having an autocorrelation function of the form $R(\tau) = \exp(-a\tau)\cos(b\tau)$. As discussed in §3.2.1(e), such a signal can be viewed as that obtained at the output of a second-order (underdamped) linear system. The transfer function equation can be written

$$\ddot{x}(t) + 2a\dot{x}(t) + (a^2 + b^2)x(t) = \sqrt{2a}\,[\dot{W}(t) + \sqrt{a^2 + b^2}\,W(t)]$$

where, again, $W(t)$ is the white noise input. The corresponding recurrence relations can be shown to be

$$x_1(i + 1) = p_1 x_1(i) + q_1 x_2(i) + \lambda_1 W(i)$$
$$x_2(i + 1) = p_2 x_1(i) + q_2 x_2(i) + \lambda_2 W(i) \tag{A.3}$$

with the required data in the x_1 array. The various parameters are given by

$$p_1 = \exp(-a\,\Delta t)\cos(b\,\Delta t) - (a/b)\sin(b\,\Delta t)$$
$$q_1 = (1/b)\exp(-a\,\Delta t)\sin(b\,\Delta t)$$
$$p_2 = [-(a^2 + b^2)/b]\exp(-a\,\Delta t)\sin(b\,\Delta t) \tag{A.4}$$
$$q_2 = \exp(-a\,\Delta t)\cos(b\,\Delta t) + (a/b)\sin(b\,\Delta t)$$

and

$$\lambda_1 = [2\sqrt{a}/(a^2 + b^2)]\{1 + \exp(-a\,\Delta t)\cos(b\,\Delta t)$$
$$+ [(\sqrt{a^2 + b^2} - a)/b]\exp(-a\,\Delta t)\sin(b\,\Delta t)\}$$
$$\lambda_2 = [2\sqrt{a}/(a^2 + b^2)]\{(2a - \sqrt{a^2 + b^2})[1 - \exp(-a\,\Delta t)\cos(b\,\Delta t)]$$
$$+ [b - a/b + (a/b)\sqrt{a^2 + b^2}]\exp(-a\,\Delta t)\sin(b\,\Delta t)\}. \tag{A.5}$$

To generate the DAT3N8 sequence the program listed below was used with $\Delta t = 0.025$, $a = 1$ and $b = \pi$. Further sequences could be generated with different dominant frequencies (set by the value of b) and/or different effective sampling rates (see §A.4). As in the previous cases, an additional scale factor is necessary to ensure a suitable standard deviation for the final sequence.

```
10 REM    ROUTINE TO CREATE A SECOND-ORDER DATA
20 REM          SET — with R(t)=Cos(b*PI*t).exp(−at)
30
40 REM    DT is the required time increment between samples
50 REM    NS is the required number of samples
60
70 VDU22,7:HIMEM=&2FFF
80 INPUT "A,B(*PI),DT,NS",A,B,DT,NS
90
100 REM    Set-up various parameters — equations A4 and A5
110
120 B=B*PI:P1I=2*PI
130 AE=EXP(−A*DT):A1=1−AE*COS(B*DT)
140 D1=SQR(A*A+B*B):D2=2*A−D1
150 B1=(A1+(D1−A)/B*AE*SIN(B*DT))*2*SQR(A)/D1
160 B2=(D2*A1+(B−A*A/B+A/B*D1)*AE*SIN(B*DT))*2
    *SQR(A)/D1
170 P1=AE*(COS(B*DT)−A/B*SIN(B*DT)):P2=−D1*D1/B*AE
    *SIN(B*DT)
180 Q1=AE/B*SIN(B*DT):Q2=AE*(COS(B*DT)+A/B
    *SIN(B*DT))
190 Y1=0:Y2=0:U1=RND(1)
200
210 REM    Recursive loop — equation A3 in Appendix
220
230 FOR I=0 TO NS−1 STEP2:U2=RND(1)
240 F=SQR(−2*LN(U1)):W1=F*COS(P1I*U2):W2=F*SIN(P1I*U2)
250 Z1=P1*Y1+Q1*Y2+B1*W1:Y2=P2*Y1+Q2*Y2+B2*W1
260 Y1=Z1:X1=160*Y1:IFABS(X1)>127THENX1=127*SGN(X1)
270 ?(&3000+I)=INT(X1+0.5)
280 Z1=P1*Y1+Q1*Y2+B1*W2:Y2=P2*Y1+Q2*Y2+B2*W2
290 Y1=Z1:X2=160*Y1:IFABS(X2)>127THENX2=127*SGN(X2)
300 ?(&3001+I)=INT(X2+0.5)
310 U1=U2:NEXT
320
330 REM    Data stored in &3000 upwards (lines 270 & 300)
```

〉 Appendix B

〉 Running the Software

〉B.1 Standard features

In this appendix the standard features of all the programs will be introduced, using the CHAP2 and CHAP4 programs as examples. It will be assumed that the reader has operating familiarity with the BBC computers and it is recommended that before doing anything else the user puts a write protect tab on the supplied system disc and makes a backup copy (both sides) on a previously formatted new disc; the latter should then be used exclusively. Note that other versions of the software may become available in future, for use with other computers.

Installing the disc and obtaining a directory will yield the list of filenames shown in table B.1. Files with the .OB extension are machine code object files, those beginning with DAT are the main data files (each holding a simulated signal of 19 456 1-byte values), those with the .M extension are small data files and the remainder are mainly BASIC files. Note that the large data files are all on side 2 of the disc so that a double-sided disc drive is essential. Alternative data files containing real (digitised) or simulated signals could be supplied by the user as required; §4.6.1 and Appendix A should provide sufficient information.

Loading is most easily done by putting the disc into drive 0, holding down the SHIFT key and pressing and then releasing the BREAK key (the so-called autoboot facility); this puts up the basic index on the screen, from which any required program can be run. Now run CHAP2. The screen should clear and then a display giving a title, a highlighted window containing instructions and a set of options, one of which is highlighted. The convention adopted

throughout the programs is that, at any point where the user has the
opportunity to change any setting, the existing setting can be retained
simply by pressing the RETURN key. In this case, therefore,
choosing one of the settings by using the cursor keys and then
pressing the RETURN key leads to the appropriate display. Further
displays under the same option (if there are any) can be obtained by
pressing the RETURN key again and the user is then prompted as
appropriate. The ESCAPE key can usually be used to return to the
main index. Similar responses are required in CHAP3.

Table B.1   Directories of both sides of the BBC disc.

Drive 0 Dir. :0.$	Option 3 (EXEC) Lib. :0.$	Drive 2 Dir. :0.$	Option 0 Lib. :0.$
!BOOT	ALL.OB	DAT0N8	DAT1N8
ALLDT.M	CHAP2	DAT2N8	DAT3N8
CHAP3	CHAP3.2		
CHAP4	CHAP4.0		
CHAP4.1	CHAP4.2		
CHAP4.3	CHAP5		
CHAP5.1	CHAP5.2		
CHAP5.3	corrraw		
DATP.OB	DNAME.B		
FFT.OB	HMOM.OB		
INDEX	PL16.OB		
PLT.OB	PRNTDMP		
PSDAT.M	RSDAT.M		
RTAU.M			

Now try the CHAP4 routines by returning to the index and running
CHAP4. Here again (and in CHAP5) displayed sets of options are
selected using the cursor keys and the RETURN key, but some
additional protocol is used. To illustrate this, choose the probability
distribution option from the CHAP4 index. The screen will first clear
and then a display appears, giving the appropriate title along with
separately highlighted windows containing the signal type (on the left)
and the printer status (on the right). The latter can be changed using
the P key and the former, like all subsequent highlighted action
prompts, can be selected using the space bar. When the user is ready
to continue, the RETURN key should be pressed, as usual.

If the printer-on option is selected, all subsequent measurement parameters and results will be sent in a suitable format to the printer; so do not choose this option if there is no printer connected! Many experiments that the reader may wish to devise will necessitate quantitative knowledge of the results, for which a hard copy is usually a great help.

With just two exceptions, all the options in CHAP4 and CHAP5 require the input of specified parameters at some point. These all have default values which can be selected simply by pressing the RETURN key. In this case (the probability distribution option) use all default values, stepping through them by using the RETURN key. After a brief calculation time, a prompt to present the results as a graph will be given and the screen will then clear, prior to a display of $p(x)$ against x (suitably normalised) appearing. For this and all other measurements which include a graphical presentation of results, it is possible to produce a disc file containing the entire screen contents, for later hardcopy on a printer using the PRNTDMP program available in the main index. Production of the disc file is initiated using the dump option—remember that the SPACE bar is used for scrolling through the available options. The following section gives further information on screen dumps.

Most of the routines in all the programs allow return to the particular chapter index by selecting the appropriate action option or using the ESCAPE key, although in some cases the latter will return you to the main index. Do not forget that the BREAK key *always* returns you to the main index.

⟩ B.2   Screen dumps

Many printers can be used to produce a screen dump, i.e. a monochrome ink image of whatever appears on the screen. Unfortunately there is no universal standard for printers or the software to drive them; so, instead of providing a direct screen-to-printer facility, the present software provides a means of copying the graphics memory to a file. The screen image can then be recreated by reading this file back into memory independently of the programs, and users can then use their own software to make a copy on a printer.

Although the details of this latter process are dependent on the hardware and operating system and will vary according to the computer model being used, many users will have access to Epson printers, or one that is software compatible. A program PRNTDMP is therefore provided under the main index; this will reload the dump file and send it to the printer. PRNTDMP is identical with that used in other volumes in this series and it has a self-explanatory dialogue. A few examples of the resulting printouts are included within the present text, as appropriate.

To produce the required file, the dump option should be selected after an action prompt, and the user is then prompted for a drive number (with the default of 2) and a file name. It should be noted that, since these files are quite long, not more than six of them can be saved on the system disc. The user may prefer to replace the latter with his or her own prepared disc, not forgetting that, before continuing to run the original program, he or she must return the system disc to drive 0. All the screen dumps are from mode 1 and BBC users may also like to be reminded that a screen dump can easily be recalled to the screen from immediate command mode, as well as from a program. For example, if DMPF1 were the name of a screen-dump file then

MODE 1:*LOAD DMPF1

would reload it.

> References

Bartlett M G 1946a *J. R. Stat. Soc. Suppl.* **8** 27

—— 1946b *J. R. Stat. Soc. Suppl.* **10** 85

—— 1955 *An Introduction to Stochastic Processes with Special Reference to Methods and Applications* 1st edn (Cambridge: CUP)

Bendat J S and Piersol A G 1966 *Measurement and Analysis of Random Data* (New York: Wiley)

Box G E P and Muller M E 1958 *Ann. Math. Stat.* **29** 610

Burg J P 1967 *37th Ann. Int. Soc. Geophys. Meet. Oklahoma City*

Fagih N 1980 *PhD Thesis* Department of Mechanical Engineering, University of Surrey

Feller W 1970 *Introduction to Probability Theory and its Applications* vol 2, 2nd edn (New York: Wiley)

Kendall M G 1973 *Time Series* (London: Charles Griffin)

Kendall M G and Stuart A 1977 *The Advanced Theory of Statistics* vols 1–3, 4th ed (London: Charles Griffin)

Lacoss R T 1971 *Geophysics* **36** 661–75

Lighthill M J 1962 *Introduction to Fourier Analysis and Generalised Functions* (Cambridge: CUP)

Miller I and Freund J E 1977 *Probability and Statistics for Engineers* (Englewood Cliffs, New Jersey: Prentice-Hall)

Priestley M B 1981 *Spectral Analysis and Time Series* (New York: Academic)

Snyder D L 1975 *Random Point Processes* (New York: Wiley)

〉 Bibliography

Barney G C 1988 *Intelligent Instrumentation* 2nd edn (London: Prentice-Hall)

Bendat J S and Piersol A G 1971 *Random Data: Analysis & Measurement Procedures* (New York: Wiley)

Betts J A 1970 *Signal Processing, Modulation and Noise* (London: Hodder & Stoughton)

Bracewell R N 1978 *The Fourier Transform and its Applications* 2nd edn (New York: McGraw-Hill)

Brigham E O 1974 *The Fast Fourier Transform* (New Jersey: Prentice-Hall)

Brook D and Wynne R J 1988 *Signal Processing: Principles and Applications* (London: Arnold)

McGillem C D and Cooper G R 1984 *Continuous and Discrete Signal & System Analysis* 2nd edn (New York: CBS College Publishing)

Meyer P L 1965 *Introductory Probability & Statistical Applications* (Reading, MA: Addison-Wesley)

Newland D E 1984 *An Introduction to Random Vibrations and Spectral Analysis* 2nd edn (New York: Wiley)

Papoulis A 1965 *Probability, Random Variables and Stochastic Processes* (New York: McGraw-Hill)

Stanley W D 1975 *Digital Signal Processing* (London: Prentice-Hall)

〉 Index

Aliasing, 84–8, 110, 112, 116, 117, 119, 130
Analogue-to-digital conversion, 1, 2, 50
Autoboot (SHIFT/BREAK), 141
Autocorrelation function
  definition, 12, 34
  computer exercises, 116–19, 130, 131
  estimation, 88–101
  examples, 35–40
  standard error, 91, 92, 95, 96
  variability, 90–5
Autocovariance, 34, 35
Autoregressive process, 40, 41, 139

Bandwidth
  amplitude, 17, 18, 48
  frequency, 35–8, 48, 73, 111
Bartlett
  estimate (spectral window), 107
  formula (autocorrelation estimation), 90–3
Bias errors, 61, 63
BREAK, 141, 143

Cardiac cycle, 45, 46
Central-limit theorem, 23, 24, 59
CHAP2, 15, 127, 141–3
CHAP3, 46, 127, 141–3
CHAP4, 76, 127–9, 141–3
CHAP5, 115, 130–2, 141–3
Chi-squared distribution, 64

Conditional sampling, 45, 46
Confidence limits, 59, 62, 63, 78–80, 128, 129
Cumulative probability distribution, 59, 60, 64, 65

Delta function, 35
Delta modulation, 49
Deterministic, 5, 6, 12, 20, 34, 43

Electrocardiogram, 45, 46
Entropy, maximum, 109, 110
Ergodicity, 8, 9
ESCAPE, 16, 142, 143
Extreme values, 33

Fourier
  fast transform (FFT), 98, 99, 116
  integral, 43
  series, 6–8, 12
  transform (discrete), 97, 99, 101

Gaussian (normal) distribution, 9, 10, 24, 33

Harmonics, 8
Higher-order moments
  computer exercises, 80, 81, 127–9
  definitions, 32, 33
  estimation, 66–8
  standard error, 67, 74
  variability, 66–8

147

Integration time, 48
Intermittent signal, 27–29

Lag time, 34, 118, 119
Linear filter, 40, 137–140

MASTER (BBC computer), 3
Mean value
  computer exercises, 78–80,
    127–9
  definitions, 30, 31
  estimation, 58–63
  standard error, 62, 63, 74
  variability, 61, 73
Mean-square value
  computer exercises, 78–80,
    127–9
  definitions, 30, 31
  estimation, 63–5
  standard error, 64, 74
  variability, 64, 65, 73
Memory, computer, 3, 133
Moving average, 41

Noise
  correlated (pink), 12, 38, 39, 44,
    137, 138
  Gaussian, 9, 10, 24, 33, 136, 137
  random, 8, 23, 32, 135, 136
  white, 10, 24, 35, 36, 44, 136,
    137
Nyquist (aliasing) frequency, 98

Oscilloscope, 4, 15

Pedestal errors, 102, 103
Periodogram, 104, 106
PRNTDMP, 129, 143, 144
Periodic signals, 5, 10–12, 20–3,
    25–7, 35, 39, 40, 44
Poisson function, 14, 39
Power spectrum, see Spectral
    density function
Probability density function
  computer exercises, 81–3, 129
  definitions, 18, 19
  estimation of, 68–71
  examples, 19–30

Probability density function (cont.)
  moments of, 17, 30, 32
  variability, 69, 74
  standard error, 69, 75
Pulse code modulation, 49

Quantisation error
  in amplitude, 24, 48–57
  in time, 84–8

Ranging errors, 51, 77
RETURN, 15, 46, 142, 143
Resolvability, 110, 113
Root-mean-square (RMS), 30, 31

Sampling
  time, 49
  rate, 49, 78, 110
SHIFT/BREAK (autoboot), 141
Simulation, signal, 133–40
Sinusoidal signals, 5, 6, 18, 20, 21,
    130
Skewness, 27, 32
SPACE, 15, 143
Spectral density function
  computer exercises, 119–23, 131,
    132
  definitions, 41, 42
  estimation of, 101–6, 110–15
  examples, 43–5
  variability, 102–6, 111
Standard deviation, 10, 31, 80
Standard error, 63, 65–7, 69, 74,
    75, 91, 92, 95, 96
Stationarity, 8, 9
Student-t distribution, 59, 60

Telegraph signal, 14, 15, 30, 32, 39
Time constant, 90, 95, 96
Truncation errors
  amplitude, 24
  autocorrelation, 107–9

Variance, 31, 63–5
Variability, 61, 63–5, 66, 69, 73,
    74, 90–5, 102–6, 111

Weiner–Khintchine theorem, 42
Windowing, 106–10

Window
  lag, 106, 107
  spectral, 106, 107